Surveys of Modern Mathematics

www.intlpress.com/smm

SMM 1: *Analytic Methods in Algebraic Geometry*
by Jean-Pierre Demailly

SMM 2: *Lie Theory and Representation Theory*
Naihong Hu, Bin Shu and Jianpan Wang, eds.

SMM 3: *Application of Elementary Differential Geometry to Influence Analysis*
Yat-Sun Poon and Wai-Yin Poon, eds.

SMM 4: *An Introduction to Rota-Baxter Algebra*
by Li Guo

SMM 6: *Open Problems and Surveys of Contemporary Mathematics*
Lizhen Ji, Yat-Sun Poon and Shing-Tung Yau, eds.

SMM 7: *Introductory Lectures on Manifold Topology: Signposts*
by Thomas Farrell and Yang Su

SMM 8: *Lie-Bäcklund-Darboux Transformations*
by Y. Charles Li and Artyom Yurov

SMM 9: *Compressible Flow and Euler's Equations*
by Demetrios Christodoulou and Shuang Miao

SMM 10: *Finite Groups: An Introduction*
by Jean-Pierre Serre

SMM 11: *Topics in Differential Geometry: Selected papers & lecture notes of Shiing-Shen Chern*
edited by Shiu-Yuen Cheng and Lizhen Ji

SMM 12: *Differential Geometry & Integral Geometry: Selected papers & lecture notes of Shiing-Shen Chern*
edited by Shiu-Yuen Cheng and Lizhen Ji

SMM 13: *Gauss–Manin Connection in Disguise: Calabi–Yau Modular Forms*
by Hossein Movasati

Surveys of Modern Mathematics

www.intlpress.com/smm

Series Editors

Surveys of Modern Mathematics
Volume 12

Differential Geometry & Integral Geometry

Selected papers & lectures of

Shiing-Shen Chern

edited by
Shiu-Yuen Cheng and Lizhen Ji

International Press
www.intlpress.com

HIGHER EDUCATION PRESS

Surveys of Modern Mathematics, Volume 12

Differential Geometry & Integral Geometry:
Selected papers & lectures of Shiing-Shen Chern

Editors:
Shiu-Yuen Cheng (Tsinghua University, Beijing)
Lizhen Ji (University of Michigan)

ISBN: 978-1-57146-342-5

Printed in the United States of America.

9 8 7 6 5 4 3 2 1

Preface

Differential geometry is a major subject in contemporary mathematics. One person who had played an essential role in the rising of differential geometry is Professor Shiing-Shen Chern.

He received the Wolf prize in 1983/4 for his "outstanding contributions to global differential geometry, which have profoundly influenced all mathematics", and the citation says:

> Professor Shiing-Shen Chern has been the leading figure in global differential geometry. His earlier work on integral geometry, especially on the kinematic formula, was the source of most later work. His groundbreaking discovery of characteristic classes (now known as Chern classes) was the turning point that set global differential geometry on a course of tumultuous development. The field has blossomed under his leadership, and his results, together with those of his numerous students, have influenced the development of topology, algebraic geometry, complex manifolds, and most recently of gauge theories in mathematical physics.

In 2001, Professor Chern received a special Morningside Lifetime Achievement Award in Mathematics, and the citation reads:

> Professor Chern is awarded the Morningside Lifetime Achievement for his work on developing the foundation of Chinese mathematics, his epochal contributions to research in differential geometry, and his nurturing of leading mathematicians both in China and abroad. In the 1940s, differential geometry was at a low point worldwide; this area of mathematics was only beginning to be understood and to be used. Professor Chern became a pioneer in this subject. Some of his major achievements include the Chern characteristic classes in fiber spaces, and his proof of the Gauss-Bonnet formula. Today, differential geometry is a major subject in mathematics and a large share of the credit for this transformation goes to Professor Chern.

One natural question is how and what did such an initiator of a major subject say about the topics dear to his heart, especially when they were taking shape. What can the young generations, especially the young Chinese mathematicians, learn from a leading figure in the last century? We hope that these two books of lecture notes of Professor Chern and some of his expository papers will answer this question.

Though there are many books now on differential geometry, integral geometry and related topics, they often lack the freshness and directness of the original descriptions by masters. This is consistent with Abel's advice "By studying the masters, not their pupils." This point was endorsed by Professor Chern in his foreword to the Chinese edition of M. Atiyah's collected works:

> No matter how refined or improved a new account is, the original papers on a subject are usually more direct and to the point. When I was young, I was benefited by the advice to read Henri Poincaré, David Hilbert, Felix Klein, Adolf Hurwitz, etc. I did better with Wilhelm Blaschke, Élie Cartan and Heinz Hopf. This has also been in the Chinese tradition, when we were told to read Confucius, Han Yu in prose, and Tu Fu in poetry.

The title of the first book *Topics in Differential Geometry* comes from the title of his lecture notes at IAS, Princeton, in 1951. It also contains his expository papers: "From Triangles to Manifolds", "Curves and Surfaces in Euclidean Space", "Characteristic Classes and Characteristic Forms", "Geometry and Physics", and "The Geometry of *G*-Structures", together with the set of so far *unpublished lecture notes*: "Minimal Submanifolds in a Riemannian Manifold" .

Together they show how differential geometry is connected to other subjects such as topology and Lie group theory. Though there are more modern expositions of these topics, they are usually not comparable with what Chern wrote. For example, his lecture notes "Topics in Differential Geometry" starts from basics which is accessible to beginners and goes right to some striking applications such as the rigidity theorem of Cohn-Vossen on convex surfaces in $\mathbb{R}^3$ and the rigidity of convex hypersurfaces in $\mathbb{R}^n$. His treatment of the theory of connection has also such characteristic of being direct and going to the key point. The papers selected to be reprinted in this book give an overview of the scope and power of differential geometry. They complement well this set of lecture notes.

This first book will be very valuable to beginners to learn the modern differential geometry, and will also be valuable to experts for them to rethink about differential geometry.

The title of the second book is a combination of titles of two sets of lecture notes of Chern which have not been formally published. It seems that there exists only one known copy of the second set of lecture notes, which is owned by the library of University of Michigan, It also contains a more recent unpublished set of lecture notes titled "Lectures on Differential Geometry".

This second book starts with a paper of Chern which gives a gentle introduction to differential geometry accessible to students and nonexperts, and a survey of the status of global differential geometry in 1971 when Chern gave a plenary talk in ICM. As Gromov commented in his Math Review of this paper:

> This is a brilliant and inspiring exposition. The author begins with a brief historical survey, outlines some fundamental notions and tools, and describes the current situation in four branches of differential geometry: manifolds of positive curvature, curvature and the Euler characteristic, minimal sub manifolds isometric mappings, holomorphic mappings.

This paper also includes some open problems and hence is also very interesting from the historical perspective.

As every student knows, there are two key components of calculus: differentiation and integration. In geometry, there are also two closely related subjects: differential geometry and integral geometry. It is valuable to read and learn them side by side.

The lecture notes "Differential Manifolds" gives a smooth and rapid introduction to differential manifolds and differential geometry. It was delivered by Chern in 1959 at University of Chicago when differential geometry was becoming a major subject in mathematics. Its freshness shines through.

As the citation of the Wolf prize indicates, Chern also made major contribution to integral geometry. The lecture notes "Lectures on Integral Geometry" give an efficient and also accessible introduction to this subject. It is worthwhile to point out that one of Chern's teachers was Blaschke, who was a major or leading figure in integral geometry. This also adds something special to his lectures. In this set of lecture notes, one can also see Chern's global view of mathematics. For example, besides standard topics in integral geometry in Euclidean spaces, it also discusses integral geometry of homogeneous spaces.

This second book will also be a very valuable introduction to both differential and integral geometry for beginners and a supplementary reading for people at all stages.

We hope and believe that there is much one can learn from these collections of lecture notes and papers of Professor Chern, a modern master in mathematics and one of the originators of global differential geometry.

In collecting material for and editing these two books, we have received generous help from Professor Chuu-Lian Terng, and support and blessing from May Chu, the daughter of Processor Chern. Liping Wang of the Higher Education Press has also been very supportive of this project. Without their kind help, these books probably cannot appear in print. We would like thank them sincerely.

Finally, may you enjoy these two books and benefit from them!

Shiu-Yuen Cheng and Lizhen Ji

January 14, 2016.

Contents

Part IV Lecture Notes on Differentiable Geometry

Part I What is Geometry and Differential Geometry

1

What Is Geometry? [1]

To avoid misunderstanding, I will not give a definition of geometry as in the customary mathematica treatment of a topic. I will only try to discuss its major historical developments.

1.1 Geometry as a logical system; Euclid

Euclid's "Elements of Geometry" (ca. 300 B.C.) is one of the great achievements of the human mind. It makes geometry into a deductive science and the geometrical phenomena as the logical conclusions of a system of axioms and postulates. The content is not restricted to geometry as we now understand the term. Its main geometrical results are:

a) Pythagoras' Theorem (Fig. 1).

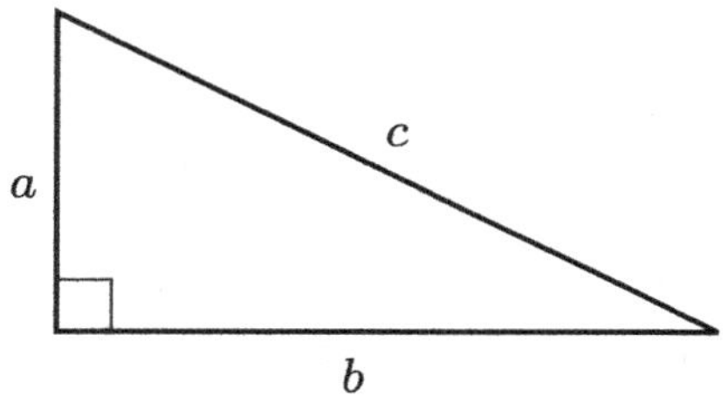

$$c^2 = a^2 + b^2$$

Fig. 1.

b) Angle-sum of a triangle (Fig. 2).

---

[1] Worked done under partial support of NSF grant DMS-87-01609. The American Mathematical Monthly, Vol. 97, No. 8, Special Geometry Issue (Oct., 1990), pp. 679–686.

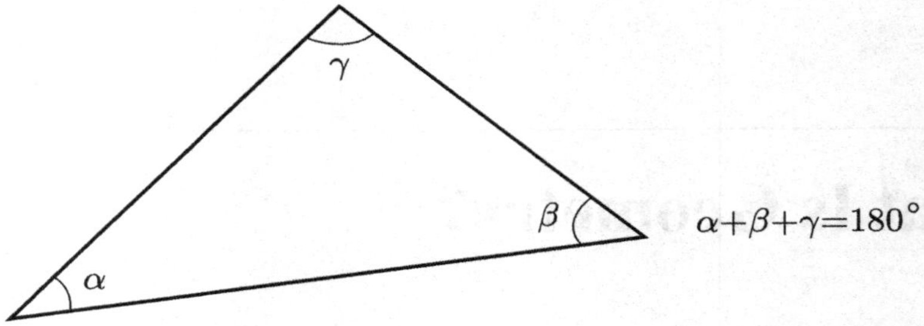

Fig. 2.

The result b) is derived using the fifth, or the last, postulate, which says: "And that, if a straight line falling on two straight lines make the angles, internal and on the same side, less than two right angles, the two straight lines, being produced indefinitely, meet on the side on which are the angles less than two right angles."(Fig. 3)

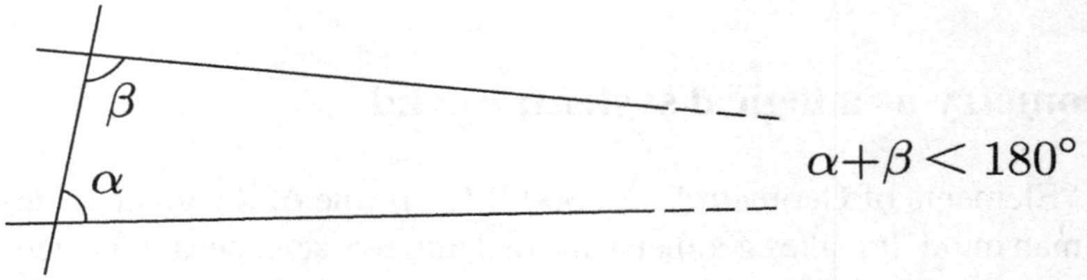

Fig. 3.

Euclid realized that the parallel postulate was not as transparent at his other axioms and postulates. Efforts were made to prove it as a consequence. Their failure led to the discovery of non-Euclidean geometry by C. F. Gauss, John Bolyai, and N. I. Lobachevski in the early 19th century.

The "Elements" treated rectilinear figures and the circles. The last three of its thirteen Books were devoted to solid geometry.

1.2 Coordinatization of space; Descartes

The introduction of coordinates by Descartes (1596–1650) was a revolution in geometry. In the plane it can be described by Fig. 4, where the role of the two coordinates x, y is not symmetric. Descartes' work was published in 1637 as an appendix, entitled "La géométrie", to his famous book on philosophy [6]. At about the same time Fermat (1601–1665) also found the concept of coordinates and used them to treat successfully geometric problems by algebraic methods. But Fermat's work was published only posthumously [7].

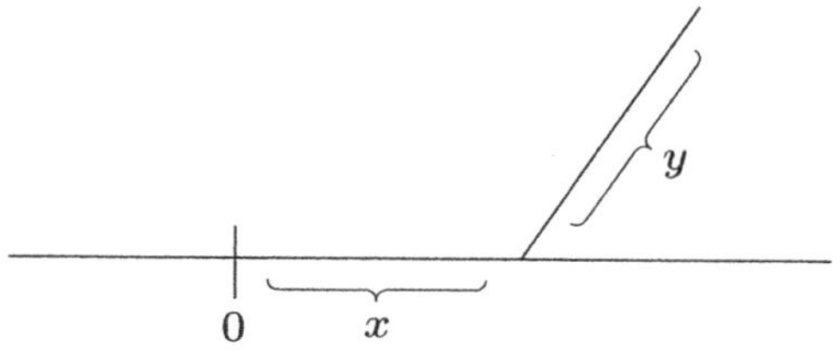

Fig. 4.

One immediate consequence was the study of curves defined by arbitrary equations

$$F(x,y) = 0, \tag{1}$$

thus enlarging the scope of the figures.

Fermat went on to introduce some of the fundamental concepts of the calculus, such as the tangent line and maxima and minima.

From two dimensions one goes to n dimensions and to an infinite number of dimensions. In these spaces one studies loci defined by arbitrary systems of equations. Thus a great vista was opened, and geometry and algebra became inseparable.

A mystery is the role of differentiation. The analytic method is most effective when the functions involved are smooth. Hence I wish to quote a philosophical question posed by Clifford Taubes [15]: Do humans really take derivatives? Can they tell the difference?

Coordinate geometry paved the way to applications to physics. An example was Newton's derivation of the Kepler laws from his law of gravitation. Kepler's first law says that the planetary orbits are ellipses with the sun as their common focus. The proof was possible only after an analytic theory of conics had been established.

1.3 Space based on the group concept; Klein's Erlanger Programm

Works on geometry led to the development of projective geometry, among whose founders wer J. V. Poncelet (1788–1867), A. F. Möbius (1790–1868), M. Chasles (1793–1880), and J. Steiner (1796–1863). Projective geometry studies the geometrical properties arising from the linear subspaces of a space and the transformations generated by projections and sections. Or the geometries resulted, the most notable ones being affine geometry and conformal geometry.

In 1872 Felix Klein formulated his Erlanger Programm [1, 11], which defines geometry as the study of the properties of a space that are invariant under a group of transformations. Thus there is a geometry corresponding to every group of transformation acting on a space. The basic notion is "group" and the notion of a space is now greatly expanded. In a certain sense the group of projective collineations is the most encompassing group and projective geometry occupies a dominant position.

The most important application of the Erlanger Programm was the treatment of non-Euclidean geometry by the so-called Cayley-Klein projective metric [12]. The hyperbolic space can be identified with the interior of a hypersphere and the non-euclidean motions with the group of projective collineations leaving invariant the hypersphere. The same group may appear as a group of transformations in different spaces. As a result the same algebraic argument could give entirely different geometric theorems. For example, everybody knows that the three medians of a triangle meet in a point. By using Study's dual numbers this translates into the following theorem of J. Petersen and F. Morley: Let ABCDEF be a skew hexagon such that consecutive sides are perpendicular. The three common perpendiculars of the pairs of opposite sides $AB, DE; BC, EF; CD, FA$ have a common perpendicular. See [13].

Sophus Lie founded a theory of general transformation groups, which became a fundamental tool of all geometry.

1.4 Localization of geometry; Gauss and Riemann

In his monograph on surface theory published in 1827 [8], Gauss (1777–1855) developed the geometry on a surface based on its fundamental form. This was generalized by B. Riemann (1826–1866) to n dimensions in his Habilitationschrift in 1854 [14]. Riemannian geometry is the geometry based on the quadratic differential form

$$ds^2 = \sum g_{ik}(u) du^i du^k, \qquad g_{ik} = g_{ki}, \qquad 1 \le i, k \le n \tag{2}$$

in the space of the coordinates $u^1, \ldots, u^n$, where the form is positive definite, or at least non-degenerate. Given ds^2, one can define the arc length of a curve, the angle between two interesting curves, the volume of a domain, and other geometrical concepts.

The main characteristic of this geometry is that it is local: it is valid in a neighborhood of the u-space. Because of this feature it fits well with field theory in physics. Einstein's general theory of relativity interprets the physical universe as a four-dimensional Lorentzian space (with a ds^2 of signature $+++-$) satisfying the field equations.

$$R_{ik} - \frac{1}{2} g_{ik} R = 8\pi\kappa T_{ik}, \tag{3}$$

where R_{ik} is the Ricci curvature tensor, R is the scalar curvature K, is a constant, and T_{ik} is the energy-stress tensor

It is soon observed that most properties of Riemannian geometry derive from its Levi-Civita parallelism, an infinitesimal transport of the tangents paces. In other words, Riemannian geometry studies the tangent bundle of a Riemannian space with the Levi-Civita connection.

1.5 Globalization; topology

Riemannian geometry and its generalizations in differential geometry are local in character. It seems a mystery to me that we do need a whole space to piece the neighborhoods together. This is achieved by topology. The notion of a differentiable manifold is one of the most sophisticated concepts of a topological manifold was made by D. Hilbert in 1902 [10, 17]. Hermann Weyl identified the Riemannian surfaces with one-dimensional complex manifolds and used it as the central theme of his epoch-making book *Die Idee der Riemannschen Fläche* [16]. On the topological side "neighborhood" became the basic concept in Hausdorff's topology [9].

Hassler Whitney saw the merit of establishing an imbedding theorem on differential manifolds (1936), thus beginning the serious study of differential topology. That derivatives plays a role in topology came as a shock when J. Milnor discovered the exotic differentiable structures on the seven-dimensional sphere (1956). By studying the Yang-Mills equations on a four-dimensional manifold, S. Donaldson found in 1983 a remarkable theorem on the intersection-form, which led to the existence of an infinite number of differentiable structures on R^4.

With the foundation of differentiable manifolds laid, geometrical structures can now be defined on them, such as the Riemannian structure, the complex structure, the conformal structure, the projective structure based on a system of paths, etc. Tools are developed for their treatment, of which the most important are the exterior differential calculus and the tensor analysis.

A fundamental notion is "curvature", in its different forms. Its simplest manifestation is the circle in plane Euclidean geometry. It could also be the force of a physical system or the strength of a gravitational or electro-magnetic field. In mathematical terms it measures the non-commutativity of covariant differentiation.

It is remarkable that suitable algebraic combinations of curvature give topological invariants. To illustrate this we wish to state the Gauss-Bonnet theorem. Let D be a domain with a sectionally smooth boundary on a two-dimensional Riemannian manifold. Then the Gauss-Bonnet theorem is the formula

$$\sum(\pi-\alpha)+\int_{\partial D} k_g ds+\int_D\int KdA = 2\pi\chi(D), \tag{4}$$

where the first term is the sum of the exterior angles at the corners, the second term is the integral of the geodesic curvature along the sides, the third term is the integral of the Gaussian curvature over D, and $\chi(D)$ is the Euler characteristic of D. For a rectilinear triangle in the Euclidean plane this is the theorem on the angle-sum stated in §1.1. For higher dimensions we will only give, for the sake of simplicity, the theorem for a compact oriented Riemannian manifold M of dimension $2n$ without boundary. Let R_{ijkl} be the Riemann-Christoffel tensor and let

$$\Omega_{ij} = \sum R_{ijkl} du^k \wedge du^l \tag{5}$$

be the "curvature form". Let

$$Pf = \sum \varepsilon_{i_1 \cdots i_{2n}} \Omega_{i_1 i_2} \wedge \cdots \wedge \Omega_{i_{2n-1} i_{2n}} \tag{6}$$

be the pfaffian, where $\varepsilon_{i_1 \dots i_{2n}}$ is $+1$ or -1 according as its indices form an even or odd permutation of $1, \dots, 2n$, and is otherwise zero, and the sum is extended over all indices from 1 to $2n$. Then the Gauss-Bonnet theorem says

$$(-1)^n \frac{1}{2^{2n} \pi^n n!} \int_M Pf = \chi(M), \tag{7}$$

where $\chi(M)$ is the Euler-Poincaré characteristic of M.

1.6 Connections in a fiber bundle; Elie Cartan

A notion which includes both Klein's homogeneous spaces and Riemann's local geometry is Cartan's generalized spaces (espaces généralisés). In modern terms it is called "a connection in a fiber bundle". It is a straightforward generalization of the Levi-Civita parallelism, which is a connection in the tangent bundle of a Riemannian manifold. In general, we have a fiber bundle $\pi : E \to M$, whose fibers $\pi^{-1}(x)$, $x \in M$, are homogeneous spaces acted on by a Lie group G. A connection is an infinitesimal transport of the fibers compatible with the group action by G.

I wish to illustrate this more precisely in the case of a complex vector bundle, where the fibers are complex vectors paces C_q of dimension q and $G = GL(q;\mathbb{C})$ [4]. The importance of complex numbers in geometry is a mystery to me. It is well organized and complete. One manifestation is the simple behaviour of the group $GL(q;\mathbb{C})$: its maximal compact subgroup $U(q)$ has no torsion and has as Weyl group the group of all permutations on q letters.

We shall call a frame n ordered set of linearly independent vectors $e_1, \dots, e_q \in \pi^{-1}(x), x \in M$. In a neighborhood U where a frame field $e_1(x), \dots, e_q(x), x \in U$, is defined, a connection is given by the infinitesimal displacement

$$De_\alpha = \sum \omega_\alpha^\beta e_\beta, \qquad 1 \le \alpha, \beta \le q, \tag{8}$$

where ω_α^β are linear differential forms in U. We call ω_α^β the connection forms and the matrix

$$\omega = (\omega_\alpha^\beta) \tag{9}$$

the connection matrix. Under a change of the frame field

$$e'_\alpha = \sum a^\beta_\alpha e_\beta, \qquad A = (a^\beta_\alpha), \tag{10}$$

the connection matrix is changed as follows:

$$\omega' A = dA + A\omega. \tag{11}$$

We introduce the curvature matrix

$$\Omega = d\omega - \omega \wedge \omega \tag{12}$$

which is a matrix of exterior two-forms. By exterior differentiation of (11) we get

$$\Omega' = A\Omega A^{-1}. \tag{13}$$

It follows that the exterior polynomial

$$\det\left(I + \frac{i}{2\pi}\Omega\right) = 1 + c_1(\Omega) + \cdots + c_q(\Omega), \tag{14}$$

in which $c_\alpha(\Omega)$ is a 2α-form, is independent of the choice of the frame field and is hence globally defined in M. Moreover, each c_α is closed, i.e.,

$$dc_\alpha = 0. \tag{15}$$

The form $c_\alpha(\Omega)$ has been called the αth Chern form of the connection and its cohomology class $\{C_\alpha(\Omega)\}$ in the sense of de Rham cohomology is an element of the cohomology group $H^{2\alpha}(M;\mathbb{Z})$ and is called the αth Chern class of the bundle E. These characteristic classes are the simplest and most fundamental global invariants of a complex vector bundle. They have the advantage of possessing a local representation, by curvature.

As in the Gauss-Bonnet formula such a representation is of great importance, because the forms $c_\alpha(\Omega)$ themselves have a geometrical significance. Moreover, let $\pi' : P \longrightarrow M$ be the bundle of frames of the complex vector bundle. Then the pull-back $\pi'^* c_\alpha$ becomes a derived form, i.e.,

$$\pi'^* c_\alpha = dTc_\alpha, \tag{16}$$

where Tc_α, a form of degree $2\alpha - 1$ in P, is uniquely determined by certain properties. This operation is called transgression and Tc_α have been called the Chern-Simons forms [5]. These formsh ave played a role in three-dimensional topology and in recent works of E. Witten on quantum field theory [20].

This theory can be developed for any fiber bundle; see [3]. The above provides the geometrical basis of gauge field theory in physics. Here M is a four-dimensional Lorentzian manifold, so that the Hodge $*$-operator is defined, and we define the codifferential

$$\delta = *d* \tag{17}$$

There is a discrepancy of terminology and notation, as given by the following table:

mathematics	physics
connection ω	gauge potential A
curvature Ω	strength F

Maxwell's theory is based on a $U(1)$-bundle over M, and his field equations can be written

$$dA = F, \qquad \delta F = J, \tag{18}$$

where J is the current vector. Actually, Maxwell wrote the first equation as

$$dF = 0, \tag{19}$$

which is a consequence. For most applications (19) is sufficient. But a critical study of an experiment proposed by Boehm and Aharanov and performed by Chambers shows that (18) are the correct equations [21]. A generalization of (18) to an $SU(2)$ bundle over M gives the Yang-Mills equation

$$DA = F, \qquad \delta F = J. \tag{20}$$

It is indeed remarkable that developments in geometry have been consistently parallel to those in physics.

1.7 An application to biology

So far the most far-reaching applications of geometry are to physics, from which it is indeed inseparable. I wish to mention an application to biology, namely, to the structures of DNA molecules. This is known to be a "double helix", which geometrically means a pair of closed curves. Their geometrical invariants will clearly be of significance in biology. The following three are most important : 1) The linking number introduced by Gauss; 2) the total twist, which is essentially the integral of the torsion; 3) the writhing number.

James White proved that between these invariants there is the relation [18]

$$Lk = Tw + Wr. \tag{21}$$

This formula is of fundamental importance in molecular biology.

1.8 Conclusion

Contemporary geometry is thus a far cry from Euclid. To summarize, I would like to consider the following as the major developments in the history of geometry:

[1] Axioms (Euclid);
[2] Coordinates (Descartes, Fermat);
[3] Calculus (Newton, Leibniz);

[4] Groups (Klein, Lie);
[5] Manifolds (Riemann);
[6] Fiber bundles (Elie Cartan, Whitney).

A property is geometric, if it does not deal directly with numbers or if it happens on a manifold, where the coordinates themselves have no meaning. Going to several variables, algebra and analysis have a tendency to be involved with geometry.

This story is clearly one-sided and incomplete, representing only my personal viewpoint, and my limitations. It is clear that the story will not end here. Recent developments in theoretical physics, such as geometric quantum field theory, string theory, etc., are pushing for a much more general definition of geometry [19].

It is satisfying to note that so far almost all the sophisticated notions introduced in geometry have been found useful.

Finally, I wish to call attention to an early paper of mine [2], which could be read as a companion to this one.

References

[1] G. Birkhoff and M. K. Bennett, Felix Klein and His "Erlanger Programm", History and Philosophy of Modern Mathematics (W. Aspray and P. Kitcher, editors), Univ. of Minn. Press, 1988, 145–176.

[2] S. Chern, From triangles to manifolds, this Monthly, 86 (1979), 339–349.

[3] S. Chern, Complex Manifolds without Potential Theory, 2nd edition, Springer, 1979.

[4] S. Chern, Vector bundles with a connection, Studies in Global Differential Geometry, Math. Asso. Amer. Studies, no. 27, (1989), 1–26.

[5] S. Chern and J. Simons, Some cohomology classes in principal fiber bundles and their application to Riemannian geometry, Proc. Nat. Acad. Sci. USA, 68 (1971), 791–794; or, characteristic forms and geometrical invariants, *Annals of Math*, 99 (1974), 48–69.

[6] René Descartes, Discours de la méthode pour bien conduire sa raison et chercher la vérité dans les sciences, 1637.

[7] Pierre de Fermat, Oeuvres, edited by Paul Tannery and Charles Henry, Gauthier–Villars, Paris, 1891–1912.

[8] C. F. Gauss, Disquisitiones generales circa superficies curvas, 1827; Ges. Werke, 4 .

[9] F. Hausdorff, Grundziüge der Mengenlehre, Leipzig 1914; dritte Auflage, Dover, N.Y. 1944; English translation, Chelsea, N.Y. 1957.

[10] D. Hilbert, Uber die Grundlagen der Geometrie, Göttinger Nachrichten, 1902, 233–241.

[11] F. Klein, Vergleichende Betrachtungen Über neuere geometrische Forschungen, Math. Annalen 43 (1893), 63–100 or Ges. Abh 1 (1921), 460–497.

[12] ________, Vorlesungen Über nicht-euklidisch Geometrie, Springer, 1928.

[13] ________, Höherve Geometrie, Springer, 1926, p. 314.

[14] B. Riemann, Über die Hypothesen welche der Geometrie zu Grunde liegen, Habilitationschrift 1854; Gött Abh 13, 1868; Ges. Werke, 1892.

[15] Clifford H. Taubes, Morse theory and monopoles; topology in longe range forces, Progress in Gauge Field Theory, Cargese 1983, 563–587, NATO Adv. Sci. Inst, Ser B, physics 115, Plenum New York-London, 1984.

[16] H. Weyl, Die Idee der Riemannschen Fläche, Leipzig, 1913; 3 te Auflage, verändert, Leipzig, 1955.

[17] ________, Riemann geometrische Ideen, ihre Auswirkung und ihre Verknüpfung mit der Gruppentheorie, Springer, 1988.

[18] James H. White, Self-linking and the Gauss integral in higher dimensions, Amer. J. Math., 91 (1969) 693–728.

[19] E. Witten, Physics and geometry, Proc. Int. Cong. of Math. Berkeley 1986, Amer. Math. Soc., 1987, Vol. 1, 267–303.

[20] ________, Quantum field theory and the Jones polynomial, Braid Group, Knot Group, and Statistical Mechanics, (C. N. Yang and M. L. Ke editors), World Scientific, 1989, 239–329.

[21] C. N. Yang, Magnetic monopoles, fiber bundles, and gauge fields, Annals of New York Academy of Sciences, 294 (1977), 86–97.

2

Differential Geometry; Its Past and Its Future [1]

2.1 Introduction

It was almost a century ago, in 1872, that Felix Klein formulated his Erlanger Programm. The idea of unifying the geometries under the group concept is simple and attractive and its applications in the derivation of different geometrical theorems from the same group-theoretic argument are usually of great elegance. This leads to the development of differential geometries of submanifolds in homogeneous (or Klein) spaces: conformal, affine, and projective differential geometries. The latter had in particular an energetic development in the twenties.

It was also about a century ago that the greatest modern differential geometer Elie Cartan was born (on April 9, 1869). Among his contributions of a basic nature are his systematic use of the exterior calculus and his clarification of the global theory of Lie groups. Fiber spaces also find their origin in Cartan's work.

Differential geometry is the study of geometry by the methods of infinitesimal calculus or analysis. Among mathematical disciplines it is probably the least understood[2]. Many mathematicians feel there is no geometry beyond two and three dimensions. The advent into higher and even infinitely many dimensions does make the intuition unreliable and the dependence on algebra and analysis mandatory. The basis of algebra is the algebraic operations and the basis of analysis is the topological structure. I would like to surmise that the core of differential geometry is the Riemannian structure (in its broad sense).

---

[1] This paper was written when the author held a research professorship of the Miller Institute and was under partial support of NSF grant GP 20096.

[2] G. D. Birkhoff, *The second is a disturbing secret fear that geometry may ultimately turn out to be no more than the glittering intuitional trappings of analysis.* Fifty years of American mathematics, *Semicentennial Addresses of Amer. Math. Soc.* (1938), p. 307.

G. W. Mackey, Geometrical intuition, while very helpful, is not reliable and cannot be depended upon for rigorous arguments, *Lectures on the Theory of Functions of a Complex Variable*, p. 21, Van Nostrand, notes.

The main object of study in differential geometry is, at least for the moment, the differentiable manifolds, structures on the manifolds (Riemannian, complex, or other), and their admissible mappings. On a manifold the coordinates are valid only locally and do not have a geometrical meaning themselves. Historically the difficulty in achieving a proper understanding of this situation must have been tremendous (I wonder whether this was part of the reason which caused Hadamard to admit his psychological difficulty in the mastery of Lie groups) [3]. For technical purposes the Ricci calculus was a powerful tool, but it is inadequate for global problems. Global differential geometry, with the exception of a few isolated results, had to wait till algebraic topology and Lie groups have paved the way.

Global differential geometry must be considered a young field. The notion of a differentiable manifold should have been in the minds of many mathematicians, but it was H. Whitney who found in 1936 a theorem to be proved: the imbedding theorem. In the case of the richer complex structure a definition of a Riemann surface by overlapping neighborhoods was given and the theory rigorously treated by H. Weyl in his famous book "Idee der Riemannschen Fläche, Göttingen, 1913" [4], following which Caratheodory gave the first definition of a high-dimensional complex manifold. More general pseudo-group structures were treated by Veblen and J. H. C. Whitehead in 1932 [34]. Only special cases of the general theory, such as Riemannian, conformal, complex, foliated structures, etc. have been found significant.

2.2 The development of some fundamental notions and tools

Perhaps the most far-reaching achievement in differential geometry in the last thirty years lies in its foundation. Not only are the notions clearly defined, but notations are devised to treat manifolds which could be infinite-dimensional. The notations are up to now on the diversive side and are thus at an experimental stage. We believe in the survival of the fittest. Important as these foundational works are, no mathematical discipline can prosper without deeper study and simple challenging problems. We will comment briefly on a few fundamental developments in differential geometry and its related subjects, without endeavoring to make the list complete.

---

[3] J. Hadamard, *Psychology of Invention in the Mathematical Field*, Princeton (1949), p. 115.

E. Cartan, in his classical *Leçons sur la géométrie des espaces de Riemann* says, *La notion générale de variété est assez difficile à définir avec précision*, p. 58.

[4] Weyl's book was dedicated to Felix Klein, to whom he acknowledged for the fundamental ideas. Weyl's definition of a Riemann surface and Hausdorff's introduction of his axioms in 1914 must have made it superfluous to give formally a definition of a differentiable manifold. Chevalley's book on Lie groups (1946) exerted a great influence in the clarification of many concepts attached to it.

(1) *Lie Groups.* — It is one of the happiest incidents in the history of mathematics that the structure of Lie groups can be so thoroughly analyzed. The existence of the five exceptional simple Lie groups makes a deep study necessary and leads to a better understanding. Even so the subject has unity and is so much simpler than (say) finite groups. The quotient spaces of Lie groups give a multitude of examples of manifolds which are easy to describe. They include the classically important spaces and form a reservoir on which new conjectures can be tested.

(2) *Fiber Spaces.* — When a manifold has a differentiable structure, it can be locally linearized, giving rise to the tangent bundle and the associated tensor bundles. The first idea of a connection in a fiber bundle with a Lie group can be found in Cartan's "espaces généralisés ". It was algebraic topology which focused on the simplest problems, e.g., the problem of introducing invariants which serve to distinguish a general fiber bundle from a product bundle. Among them are the characteristic classes. Characteristic classes with real coefficients can be represented by the curvature of a connection, the simplest example being the Gauss-Bonnet formula. The bundle structure is now an integral part of differential geometry.

(3) *Variational Methods.* — The importance of the notion of measure (length, area, volume, curvature, etc.) makes the variational method a powerful and indispensable tool. The study of geodesies on a Riemannian manifold is a brilliant chapter of mathematics. It led to Morse's creation of the critical point theory whose scope extends far beyond differential geometry. Another example is the Dirichlet problem and its application to elliptic operators. Multiple integral variational problems open a vista whose terrain is still rocky. It promises, however, a fertile field of work. When a geometrical problem involves a function, either over the given manifold or in some related functional space, it always pays to look at its critical values and the second variation at them. Much of differential geometry utilizes this idea, in its various ramifications. The importance of variational method in differential geometry can hardly be over-emphasized.

(4) *Elliptic Differential Systems.* — The geometrical properties of differential geometry are generally expressed by differential equations or inequalities. Contrary to analysis special systems with their special properties received more attention. While analysis is the main tool, geometry furnishes the variety. Differential systems on manifolds with or without boundary are the prime objects of study.

Elliptic systems occupy a central position because of their rich properties, which follow from the severe restrictions on the set of solutions. Hodge's harmonic differential forms, with their applications to Kahlerian manifolds, will remain a crucial landmark. A simple idea of Bochner relates them to curvature and leads to vanishing theorems when the curvature satisfies proper "positivity" conditions. This has remained a standard method in the establishment of such theorems, which in turn give rise to existence theorems. The indices of linear elliptic operators on a compact manifold include some of the deepest invariants of manifolds (Atiyah, Bott, Singer).

In the study of mappings an important problem consists in the analysis of the singularities. Important progress has been made recently on the singularities of differentiable mappings (Whitney, Thom, Malgrange, Mather). If the mappings are defined by elliptic differential equations, cases are known where the singularities take relatively simple form. Singularities in differential geometry remain a relatively untouched subject.

2.3 Formulation of some problems with discussion of related results

We will attempt to discuss some areas where it is believed that fruitful researches can be carried out. The limited time at my disposal and, above all, my own limitation make it impossible for the treatment to be even remotely exhaustive. Any subject left out carries no implication that it is considered less significant.

My object is to amuse you by stating some very simple problems which have so far defied the efforts of geometers. The danger in formulating such problems is that the line distinguishing them from mathematical puzzles is thin. Personally I think there is no such line except that the "serious" problems concern with a new domain where the phenomena are not well understood and the basic concepts not well developed. Geometry and analysis on manifolds are still at this stage and will remain so for years to come. When such problems are solved, similar ones will tend toward puzzles. Many of the problems to be given below are known. It is hoped that its collection may attract mathematicians not engaged in this field and lead to further progress.

2.3.1 Riemannian manifolds whose sectional curvatures keep a constant sign

It was known to Riemann that the local properties of a Riemannian structure are completely determined by its sectional curvature. The latter is a function $R(\sigma)$ of a two-dimensional subspace σ of the tangent space at a point x, which is equal to the gaussian curvature of the surface generated by the geodesies tangent to σ at x. Manifolds for which $R(\sigma)$ keeps a constant sign for all σ have a simple geometrical meaning. For their global study it is important to require that they are not proper open subsets of larger manifolds and, following Hopf and Rinow, it is customary to impose the stronger completeness condition: every geodesic can be indefinitely extended. In fact, without the completeness requirement the sign of the sectional curvature imposes hardly any condition on the manifold, as Gromov [21] proved that there exists on any non-compact manifold a Riemannian metric for which the range of the values of $R(\sigma)$ is any open interval on the real line.

For complete Riemannian manifolds M for which $R(\sigma)$ keeps the same sign the two classical theorems are:

(1) THEOREM OF HADAMARD-CARTAN. — If $R(\sigma) \leq 0$, the universal covering manifold of M is diffeomorphic to $R^n, n = \dim M$.

(2) THEOREM OF BONNET-MYERS. — If $R(\sigma) \geq c$ (= const) > 0, M has a diameter $\leq \pi/c^{1/2}$ and is therefore compact.

The case of positive curvature turns out to be more elusive. Cheeger and Gromoll [9] achieved what is essentially a structure theory of non-compact complete Riemannian manifolds M with $R(\sigma) \geq 0$ (all σ) by proving the following theorem. There is in M a compact totally geodesic and totally convex submanifold S_M (to be called the soul of M) without boundary such that M is diffeomorphic to the normal bundle of S_M. If the sectional curvature is strictly positive, then Gromoll and Meyer [20] proved that the soul is a point and M is diffeomorphic to R^n. In particular, M must be simply connected.

Compact Riemannian manifolds of positive curvature obviously satisfy the stronger condition $R(\sigma) \geq c > 0$ (all σ). By the Bonnet-Myers Theorem they are identical with the complete Riemannian manifolds with the same property. They are not necessarily simply connected, as the example of the non-euclidean elliptic space shows. So far the simply connected compact differentiable manifolds known to admit a Riemannian metric of positive curvature are the following [3]: (1) the n-sphere; (2) the complex projective space; (3) the quaternion projective space; (4) the Cayley plane; (5) two manifolds discovered by Berger, of dimensions 7 and 13 respectively.

It is very unlikely that there are no others, but nothing more is known. The following question was asked by H. Hopf:

Problem 2.3.1. Does the product of two 2-dimensional spheres admit a Riemannian metric of strictly positive curvature?

More generally, it is not known whether the exotic 7-spheres, some of which are bundles of 3-spheres over 4-spheres, admit Riemannian metrics of positive curvature. The answer to the question in Problem 2.3.1 is probably negative. A supporting evidence is furnished by the following theorem of Berger [5] : Let M and N be compact Riemannian manifolds. Let $g(t)$ be a family of Riemannian structures on $M \times N$, such that $g(0)$ is the product structure and such that the following condition is satisfied:

$$\left. \frac{dR(\sigma)}{dt} \right|_{t=0} \geq 0$$

for all σ spanned at $x \in M \times N$ by a tangent vector to M and a tangent vector to N. Then

$$\left. \frac{dR(\sigma)}{dt} \right|_{t=0} = 0$$

for all such σ.

To get deeper topological properties of a manifold of positive curvature Rauch introduced the notion of pinching. M is said to be β-pinched if $0 < \beta < R(\sigma) \leq 1$ for all σ. After the pioneering work of Rauch the following are the main theorems on the topology of compact pinched Riemannian manifolds of positive curvature:

(1) (Berger-Klingenberg) [4, 25]. If a simply connected Riemannian manifold of positive curvature is β-pinched, $\beta > \frac{1}{4}$, it is homeomorphic to the n-sphere; if $\beta = \frac{1}{4}$ and it is not homeomorphic to the n-sphere, it is isometric to a symmetric space of rank 1.

(2) (Gromoll-Calabi) [19]. Let M be an n-dimensional compact simply connected Riemannian manifold of positive curvature. There exists a universal constant $\beta(n) < 1$, depending only on n, such that if M is $\beta(n)$-pinched, it is diffeomorphic to the standard n-sphere.

Similar problems can be studied on the global implications of curvature properties of complex Kählerian manifolds. A new feature is the notion of holomorphic sectional curvature, i.e., sectional curvature $R(\sigma)$, where σ is the two-dimensional real space underlying a complex line in the complex tangent space. A most attractive question is the following one formulated by Frankel:

Problem 2.3.2. Let M be a compact Kählerian manifold of positive sectional curvature. Is M biholomorphically equivalent to the complex projective space?

Andreotti and Frankel [17] proved that the answer is affirmative if M is of dimension 2. The proof makes use of the classification of algebraic surfaces. Partial results were recently obtained by Kobayashi and Ochiai [26] for 3 dimensions.

2.3.2 Euler-Poincaré characteristic

Among the important topological invariants of a manifold is the Euler-Poincaré characteristic. Its role is well-known on problems such as the Lefschetz fixed-point theorem, singularities of vector fields, and indices of some elliptic operators. Geometrically it is closely related to the total curvature (*curvatura integra*) as expressed by the Gauss-Bonnet formula

$$\chi(M) = \frac{(-1)^m}{2^{3m}\pi^m m} \int_M \left(\sum_{i,j} \varepsilon_{i_1 \dots i_{2m}} \varepsilon_{j_1 \dots j_{2m}} R_{i_1 i_2 j_1 j_2} \cdots R_{i_{2m-1} i_{2m} j_{2m-1} j_{2m}} \right) dv. \quad (1)$$

Here M is a compact orientable Riemannian manifold of even dimension $n = 2m$, $\chi(M)$ is its Euler-Poincaré characteristic, dv is the volume element, and R_{ijkl} are the components of the curvature tensor relative to orthonormal frames. The $\varepsilon_{i_1 \dots i_{2m}}$ is the Kronecker symbol and is zero if $i_1, \dots, i_{2m}$ do not form a permutation of 1, $\dots$, $2m$ and is equal to $+1$ or -1 according as the permutation is even or odd.

In spite of the explicit expression for $\chi(M)$ the following has not been established:

Problem 2.3.3. AND CONJECTURE. — If M has sectional curvatures ≥ 0, then $\chi(M) \geq 0$. If M has sectional curvatures ≤ 0, then $\chi(M) \geq 0$ or ≤ 0, according as $n \equiv 0$ or $2 \bmod 4$.

The above statement has been proved for $n = 4$ [10] and for the case that M has constant sectional curvature. A first approach would be to study the sign of the integrand in the Gauss-Bonnet formula, a purely algebraic problem. Even this algebraic problem seems to be of great interest [33].

As with the classical Gauss-Bonnet formula the relationship is more useful for compact manifolds with boundary (in which case a boundary integral should be added to make the formula (1) valid) and the problem is more interesting for non-compact manifolds, because a deeper study of the geometry will then be necessary. We will denote by $C(M)$ the right-hand side of (1) and we shall formulate the problem:

Problem 2.3.4. Let M be a complete Riemannian manifold of even dimension. Suppose $\chi(M)$ and $C(M)$ both exist, the latter meaning that the corresponding integral converges. Find a geometrical interpretation of the difference

$$\delta(M) = \chi(M) - C(M).$$

Of course $\delta(M) = 0$ if M is compact. In two dimensions Cohn-Vossen's classical inequality says that $\delta(M) \geq 0$. For a class of two-dimensional manifolds Finn and A. Huber [16, 23] obtained a geometrical interpretation of $\delta(M)$, which implies that it is non-negative. Partial results on Problem 2.3.4 have been obtained by E. Portnoy [30]. Perhaps the case of Kählerian manifolds has a simpler answer and should be studied first.

In a different direction Satake [31] obtained a Gauss-Bonnet formula for his V-manifolds and applied it to automorphic functions and number theory. V-manifolds are essentially manifolds with singularities of a relatively simple type.

Another problem on the Euler-Poincaré characteristic concerns compact affinely connected manifolds which are locally flat. These can be described as manifolds with a linear structure, i.e., having a covering by coordinate neighborhoods such that the coordinate transformation in overlapping neighborhoods is linear.

Problem 2.3.5. Let M be a compact manifold with an affine connection which is locally flat. Is its Euler-Poincarés characteristic equal to zero?

Bensecri proved that the answer is affirmative if M is of two dimensions (for proof and generalization cf. Milnor [27]). The high-dimensional case has been investigated by L. Auslander who proved the theorem [1]: suppose the affine connection be complete and suppose that the homomorphism $h : \pi_1(M) \to GL(n,R)$ defined by the holonomy group is not an isomorphism of the fundamental group $\pi_1(M)$ onto a discrete subgroup of $GL(n,R)$. Then $\chi(M) = 0$.

It is not known whether h can imbed $\pi_1(M)$ as a discrete subgroup of $GL(n,R)$.

In spite of great developments in algebraic topology there are simple problems on the Euler-Poincaré characteristic which remain unanswered.

2.3.3 Minimal submanifolds

A minimal submanifold is an immersion $x : M^n \longrightarrow X^N$ of an n-dimensional differentiable manifold M'' (or simply M) into a Riemannian manifold X^N of dimension N, which locally solves the Plateau problem: Every point $x \in M$ has a neighborhood U such that U is of smallest n-dimensional area compared with other n-dimensional submanifolds having the same boundary ∂U. Analytically the condition can be expressed as follows: Let D^2x be the second differential on M in the sense of Levi- Civita. Then (D^2x, ξ), where ξ is a normal vector to M at x, is a quadratic differential form, the second fundamental form relative to ξ. The differential equation to be satisfied by M is

$$\mathrm{Tr}(D^2x, \xi) = 0, \qquad \text{all } \xi. \tag{2}$$

It is a system of non-linear elliptic partial differential equations of the second order, whose number is equal to the codimension $N - n$. A minimal submanifold of dimension one is a geodesic.

We wish to study the properties of complete minimal submanifolds in a given Riemannian manifold X^N (cf. [12]). Except for geodesies the interest has so far been restricted to the case when the ambient space X^N is either the Euclidean space E^N or the unit sphere $S^N(1)$ imbedded in E^{N+1}.

For a minimal submanifold $x : M^n \longrightarrow E^N$ in the Euclidean space a condition equivalent to (2) is that the coordinate functions are harmonic (relative to the induced metric). It follows that for $n > 0$ a complete minimal submanifold in E^N is noncompact.

For various reasons the case of codimension one (i.e., the minimal hypersurfaces) is the most important. Let $x_1, \ldots, x_n, z$ be the coordinates in E^{n+1}. Consider minimal hypersurfaces defined by the equation

$$z = F(x_1, \ldots, x_n) \tag{3}$$

for all $x_1, \ldots, x_n$. The following fundamental theorem generalizes the classical theorem of Bernstein and was the combined effort of de Giorgi ($n = 3$), Almgren ($n = 4$), Simons ($n \leq 7$), Bombieri, de Giorgi, Giusti ($n \geq 8$) [6, 32]. The minimal hypersurface defined by (3) must be a hyperplane for $n \leq 7$ and is not always a hyperplane for $n \geq 8$.

The main reason for this difference is the existence of absolute minimum cones in high-dimensional Euclidean space, which in turn depends on properties of compact minimal hypersurfaces in $S^N(1)$. From a general viewpoint the study of compact minimal submanifolds in $S^N(1)$ is attractive for its own sake. The first uniqueness theorem is the theorem of Almgren-Calabi [11]. If a two-sphere is immersed as a minimal surface in $S^3(1)$, it must be the equator.

By a counter-example of Hsiang [22] this theorem is not true for the next dimension. However, the following question, which can be designated as the "spherical Bernstein problem", is unanswered:

Problem 2.3.6. Let the n-sphere be imbedded as a minimal hypersurface in $S^{n+1}(1)$. Is it an equator?

Two-dimensional minimal surfaces in E^N and in $S^N(1)$ have been more thoroughly studied, because of the application of complex function theory. If the surface is itself a two-sphere (hence in $S^N(1)$), severe restriction is imposed for global reason and we have the following theorem (Boruvka, do Carmo, Wallach, Chern, but mainly Calabi [8, 14]). Let the two-sphere be immersed in $S^N(1)$ as a minimal surface, such that it does not belong to an equator. Then we have: (1) N is even; (2) The total area of the surface is an integral multiple of 2π; (3) If the induced metric is of constant Gaussian curvature, it is completely determined up to motions in $S^N(1)$ and the Gaussian curvature has the value

$$K = \frac{2}{m(m+1)}, \qquad N = 2m. \tag{4}$$

(4) There are minimal two-spheres in $S^N(1)$ of non-constant Gaussian curvature; all these with a given area form a finite-dimensional space.

The immersion of the n-sphere as a minimal submanifold of $S^N(1)$ is a fascinating problem. If the induced metric has constant curvature, the immersion is given by the spherical harmonics (Takahashi). For $n > 2$ two isometric minimal immersions $S^n(a) \to S^N(1)$ are not necessarily equivalent under the motions of the ambient space (do Carmo, Wallach [15]). In view of the precise results on the two-sphere we wish to propose the following problem:

Problem 2.3.7. Consider minimal immersions $S^n \to S^N(1)$ with total area $\leq A$ (= const) and identify those which differ by a motion of the ambient space. Is the resulting set a finite-dimensional space with some natural topology?

2.3.4 Isometric mappings

A differentiable mapping $f: M \to V$ of Riemannian manifolds is called isometric if it preserves the lengths of tangent vectors. It is therefore necessarily an immersion, and $\dim M \leq \dim V$. Classical differential geometry deals almost exclusively with the case that V is the Euclidean space E^N of dimension N. We believe this is the most interesting case and we will adopt this restriction in our discussion.

The first problem is that of existence. Since the fundamental tensor on a Riemannian manifold of dimension n involves $n(n+1)/2$ components, Schläfli conjectured in 1871 that every Riemannian manifold of dimension n can be locally imbedded in E^N, with $N = \frac{1}{2}n(n+1)$. This was proved by Elie Cartan in 1927 for the real analytic case. For smooth non-analytic manifolds this local isometric imbedding problem is unsolved, even for $n = 2$, unless some restriction on the metric is imposed such as the Gaussian curvature keeping a constant sign. In other words, it is not known whether any smooth two-dimensional Riemannian manifold can be locally isometrically imbedded in E^3. The answer is probably negative.

The two important global imbedding theorems are:

(1) (Weyl's Problem). A compact two-dimensional Riemannian manifold of positive Gaussian curvature can be isometrically imbedded in E^3 (as a convex surface).

(2) (Nash's Theorem [18, 28]). A compact (resp. non-compact) C^∞ Riemannian manifold of dimension n can be isometrically imbedded in E^N,

$$N = \frac{1}{2}n(3n+11) \quad (\text{resp. } N = 2(2n+1)(3n+7)).^5$$

The second problem is the uniqueness of the isometric imbedding, also called rigidity, which is the problem whether an isometric immersion is determined up to a rigid motion of the ambient space E^N. Most interesting is the classical case of surfaces in E^N. Cohn-Vossen proved the rigidity of compact surfaces with Gaussian curvature $K > 0$ and the theorem was extended by Voss [35] to the case $K \geq 0$. Even before Cohn-Vossen, Liebmann proved that a smooth family of isometric compact convex surfaces (i.e., $K > 0$) is trivial, i.e., it consists of the surfaces obtained by the rigid motion of one member of the family. It is not known whether the same is true when the curvature condition is dropped and we believe the following problem is fundamental:

Problem 2.3.8. Let M be a compact surface and I be the interval $-1 < t < 1$. Let $f : M \times I \to E^3$ be a differentiable mapping such that $f_t : M \to E^3$ defined by $f_t(x) = f(x,t)$, $x \in M$, $t \in I$, is an immersion for each t. Suppose that the metric ds_t^2 induced by f_t on M is independent of t. Does there exist a rigid motion $g(t)$ such that

$$f_t(x) = g(t) f_0(x), \qquad x \in M, \tag{5}$$

where the right-hand side denotes the action on f_0 by $g(t)$?

The following remarks may be relevant to the problem. Cohn-Vossen [13] proved the existence of an unstable family of compact surfaces of revolution, i.e., that the above conclusion is not true if the hypothesis that ds_t^2 is independent of t is replaced by

$$\frac{\partial}{\partial t} ds_t^2|_{t=0} = \frac{\partial^2}{\partial t^2} ds_t^2|_{t=0} = 0. \tag{6}$$

There are well-known examples showing that Cohn-Vossen's rigidity theorem is not true without the convexity condition $K \geq 0$. A generalization of the latter condition to surfaces of higher genus is the notion of tightness. Let $f : M \to E^3$ be an immersed surface. The tangent plane at a point x is a local (resp. global) support plane if a neighborhood of the surface at x (resp. the whole surface $f(M)$) lies at one side of it. The surface is called tightly immersed if every local support plane

5 The value for N in the case of non-compact manifolds is an improvement of Nash's value by Green [18].

is a global support plane. A. D. Alexandrow proved that a real analytic tightly imbedded surface of genus one is rigid and Nirenberg [29] replaced the analyticity condition by some other conditions.

On the other hand, the notion of tightness has a meaning for polyhedral surfaces. In this case the rigidity problem asks whether the congruence of corresponding faces of two tightly imbedded polyhedral surfaces implies that they differ by a rigid motion. Cauchy's classical theorem says that this is true if the surfaces are of genus zero. But Banchoff [2] has constructed examples showing that this is untrue for surfaces of genus one. From these remarks it is anybody's guess whether the answer to the question in Problem 2.3.8 is affirmative or negative.

When M is of dimension greater than two, isometry is a strong condition and there are local rigidity theorems.

2.3.5 Holomorphic mappings

A holomorphic mapping $f: M \to V$ of complex manifolds is a continuous mapping which is locally denned by expressing the coordinates of the image point as holomorphic functions of those of the original point. The most significant example is the case when M is the complex line C and V is the complex projective line $P_1(C)$ (or the Riemann sphere), in which case the mapping is known as a meromorphic function. Much recent progress has been made in extending classical geometrical function theory to the study of holomorphic mappings.

A holomorphic mapping is called non-degenerate if the Jacobian matrix is of maximum rank at some point. For given M, V there may not exist a non-degenerate holomorphic mapping. Let B be a closed subset of V. Classically the following problem has been much studied.

Intersection or non-existence problem. Find B such that there is no non-degenerate holomorphic mapping $M \to V - B$, i.e., every non-degenerate holomorphic mapping $f: M \to V$ has the property $f(M) \cap B \neq \emptyset$.

The Picard theorem concerns the case $M = C, V = P_1(C)$, and B is the set of three distinct points. Clearly if the property holds for B, it holds for a subset containing B, so that a stronger theorem results from a smaller subset B. In view of the extreme importance and elegance of the Picard theorem, we wish to state the following conjecture of Wu:

Problem 2.3.9. Let C_n be the n-dimensional complex number space and $P_n(C)$ the n-dimensional complex projective space. Let B be the set of $n+2$ hyperplanes of $P_n(C)$ in general position (i.e., any $n+1$ of them are the faces of a non-degenerate n-simplex). Then there is no non-degenerate holomorphic mapping $C_n \to P_n(C) - B$.

The Picard theorem says that this is true for $n = 1$. Wu has established this for $n \leq 4$. Moreover, if we set

$$\rho(n) = \begin{cases} \left(\frac{n}{2}+1\right)^2+1, & n \text{ even}, \\ \left(\frac{n+1}{2}\right)\left(\frac{n+3}{2}\right)+1, & n \text{ odd}, \end{cases}$$

and let B' be the set of $\rho(n)$ hyperplanes in general position in $P_n(C)$, then Wu [36] proved that every holomorphic mapping $f: C_n \to P_n(C) - B'$ must reduce to a constant.

A far-reaching generalization of the Picard theory is the equi-distribution theory of Nevanlinna, which studies the frequency that a non-constant meromorphic function takes given values. In terms of vector bundles the problem can be generalized as follows [7]. Let M be a complex manifold and $p : E \to M$ a holomorphic vector bundle over M. A holomorphic mapping $s : M \to E$ is called a section if $p \cdot s =$ identity. Let W be a finite-dimensional vector space of holomorphic sections. Suppose the manifold and the bundle fulfill some convexity conditions (which are automatically satisfied in the classical case). Then we can define, to each $s(\neq 0) \in W$, a defect $\delta(s)$ satisfying the conditions: (1) $0 \leq \delta(s) < 1$; (2) $\delta(\lambda s) = \delta(s)$, $\lambda \in C - \{0\}$; (3) $\delta(s) = 1$ if s has no zero. The equi-distribution problem is to find an upper bound of an average of $\delta(s)$ (a sum in the case of a finite number of sections and an integral in the case of an infinite set). The problem has been studied recently by several authors.

Dual to the intersection problem is the extension problem: Given complex manifolds M, V and a closed subset $A \subset M$. When is a holomorphic mapping $M - A \to V$ the restriction of a holomorphic mapping $M \to V$?

Many extension theorems are known. In several complex variables the most famous are the Hartogs and Riemann extension theorems, which concern with the case that V is either the complex line or a bounded set of it. We wish to formulate the following problem of Hartogs type where the curvature of the image manifold enters into play:

Problem 2.3.10. Let Δ be an n-ball in $C_n, n \geq 2$, and let V be a complete hermitian manifold of holomorphic sectional curvature ≤ 0. Is it true that every holomorphic mapping of a neighborhood of the boundary $\partial\Delta$ of Δ into V extends into a holomorphic mapping of Δ into V?

It is known that without the curvature condition on V the assertion is not true [24]. The problem belongs to an area which might be described as "hyperbolic complex analysis". The philosophy is that negative curvature of the receiving space limits the holomorphic mappings and allows strong theorems. In fact, a bounded holomorphic function is a mapping into a ball which has the non-euclidean hyperbolic metric.

References

[1] L. Auslander, The structure of complete locally affine manifolds, *Topology*, 3 (1964), suppl. 1, pp. 131–139.

[2] T. Banchoff, Non-rigidity theorems for tight polyhedra, to appear in *Archiv d. Math.*

[3] M. Berger, Les variétés riemanniennes homogènes normales simplement connexes a courbure strictement positive, *Ann. Scuola Norm. Sup. Pisa*, 15 (1961), pp. 179–246.

[4] ________, Sur quelques variétés Riemanniennes suffisamment pincées, *Bull. Soc. Math. France*, 88 (1960), pp. 57-71.

[5] ________, Trois remarques sur les variétés riemanniennes à courbure positive, *Comptes Rendus* (Paris), 263 (1966), A 76–A 78.

[6] Bombieri, E. De Giorgi and E. Giusti, Minimal cones and the Bernstein problem, *Inventiones Math.*, 1 (1969), pp. 243–268.

[7] R. Bott and S. S. Chern, Hermitian vector bundles and the equidistribution of the zeroes of their holomorphic sections, *Acta Math.*, 114 (1965), pp. 71–112.

[8] E. Calabi, Minimal immersions of surfaces in euclidean spheres, */. Diff. Geom.*, I (1967), pp. 111–125.

[9] J. Cheeger and D. Gromoll, The structure of complete manifolds of non-negative curvature, *Bull. Amer. Math. Soc.*, 74 (1968), pp. 1147–1150.

[10] S. Chern, On curvature and characteristic classes of a Riemann manifold, *Abh. Math. Sem. Univ. Hamburg*, 20 (1955), pp. 117–126.

[11] ________, Simple proofs of two theorems on minimal surfaces, *L'Enseig. Math.*, 15 (1969), pp. 53–61.

[12] ________, Brief survey of minimal submanifolds, *Tagungsberichte, Oberwolfach* (1969).

[13] S. Cohn-Vossen, Unstarre geschlossene Flächen, *Math. Ann.*, 102 (1929), pp. 10–29.

[14] M. Do Carmo and N. Wallach, Representations of compact groups and minimal' immersions into spheres, to appear in *J. Diff. Geom.*

[15] ________, Minimal immersions of spheres into spheres, *Proc. Nat. Acad. Sci* (USA), 63 (1969), pp. 640–642.

[16] R. Finn, On a class of conformal metrics, with application to differential geometry in the large, *Comm. Math. Helv.*, 40 (1965), pp. 1–30.

[17] T. Frankel, Manifolds with positive curvature, *Pacific J. Math.*, 11 (1961), pp. 165–174.

[18] Robert E. Greene, Isometric embeddings of riemannian and pseudo-riemannian manifolds, *Memoirs Amer. Math. Soc.*, No. 97 (1970).

[19] D. Gromoll, Differenzier bare Strukturen und Metriken positiver Krümmung auf Spharen, *Math. Annalen*, 164 (1966), pp. 353–371.

[20] D. Gromoll and W. Meyer, On complete open manifolds of positive curvature, *Annals of Math.*, 90 (1969), pp. 75–90.

[21] M. L. Gromov, Stable mappings of foliations into manifolds, *Izvestia Akad. Nauk SSSR, Ser. Mat.*, 33 (1969), pp. 707–734.

[22] W. Y. Hsiang, Remarks on closed minimal submanifolds in the standard riemannian m-sphere, *J. Diff. Geom.*, 1 (1967), pp. 257–267.

[23] A. Huber, On subharmonic functions and differential geometry in the large, *Comm. Math. Helv.*, 32 (1957), pp. 13–72.

[24] H. Kerner, Über die Fortsetzung holomorpher Abbildungen, *Archiv d. Math.*, 11 (1960), pp. 44–47.

[25] W. Klingenberg, Über Riemannsche Mannigfaltigkeiten mit nach oben beschränkter Krümmung, *Annali di Mat.*, 60 (1962), pp. 49–59.

[26] S. Kobayashi and T. Ochiai, On complex manifolds with positive tangent bundles, to appear in *J. Math. Soc. Japan*.

[27] J. Milnor, On the existence of a connection with curvature zero, *Comm. Math. Helv.*, 32 (1958), pp. 215–223.

[28] J. Nash, The imbedding problem for riemannian manifolds, *Ann. of Math.*, 63 (1956), pp. 20–64.

[29] L. Nirenberg, Rigidity of a class of closed surfaces, Nonlinear Problems, R. E. Langer (Editor) (1963), pp. 177–193.

[30] E. Portnoy, Toward a generalized Gauss-Bonnet formula for complete open manifolds, Stanford thesis, (1969).

[31] I. Satake, The Gauss-Bonnet theorem for V-manifolds, *J. Math. Soc. Japan*, 9 (1957), pp. 464–492.

[32] J. Simons, Minimal varieties in riemannian manifolds, *Ann. of Math.*, 88 (1968), pp. 62–105.

[33] J. Thorpe, The zeroes of non-negative curvature operators, to appear.

Part II Lectures on Integral Geometry

3

Lectures on Integral Geometry

Introduction

The eight lectures contained in these notes may be divided into two parts. The first four lectures treat plane integral geometry, while the remaining four lectures concern the integral geometry in a higher dimensional space. In the last lecture, several detailed discussions and proofs are omitted; the interested reader can find these materials in the original papers indicated.

Professor Chern would like to apologize that he did not have time to reread the manuscript. On account of this fact, all mathematical and grammatical errors, inclarities, illogical arrangements etc., that exist in this book should be interpreted as the faults caused by the incapability on the part of the undersigned, who is solely responsible to them.

3.1 Lecture I

I shall devote all my efforts in the discussions of Integral Geometry this time. This is an important branch in geometry which has some meaningful developments in recent years. Many papers appear recently which concern different aspects of this branch. The first few lectures of mine will be very elementary, and most of the materials can be found in the book: L.A. Santalo, *Introduction to Integral Geometry*, Paris, 1953.

After these fundamental preparations we shall come into a discussion of its important developments.

The old integral geometry comes from the calculations of geometric probabilities. The simplest examples are the following:

3.1.1 Buffon's needle problem

Let the plane be equipped with a system of parallel straight lines such that every pair of consecutive straight lines has a constant distance a, and we are given a needle of length ℓ. (Let us suppose $\ell < a$.) Now throw the needle on the plane, what is the probability that it will meet one of the parallel lines?

The answer, which can be obtained from a simple calculation (cf. Gnedenko, Theory of Probability, p. 37), is $\frac{2\ell}{\pi a}$.

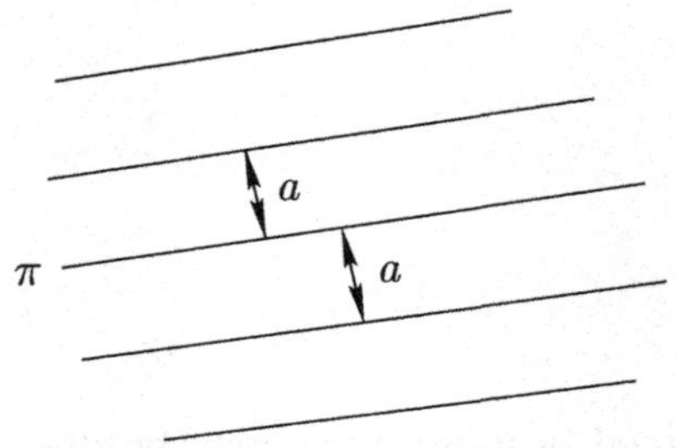

Fig. 1.

This result can also be deduced from a more general theorem which will be treated later.

In the case when $\ell > a$, the same problem still exists, and it can actually also be solved, but it is more complicated.

There was once a time when people evaluated the number π through examinations by using the above formula $\frac{2\ell}{\pi a} =$ probability.

In fact, the results of their works have proved to be very accurate.

The second example I want to talk about is the so-called:

3.1.2 Bertrand's parabox

Cut the unit circle by a line, what is the probability that length of chord is greater than $\sqrt{3}$, which is the length of one side of an inscribed, equilateral triangle. Now Bertrand gave three different solutions to this problem.

First solution: A chord can be determined by fixing the origin O on the circumference of the unit circle and giving the two arcs α, β counted from O to the points of intersection. By symmetry, we can suppose $\alpha = O$, i.e. the chord pass through O.

Draw the equilateral inscribed Δ having one vertex at O. It is obvious that a chord OA is of length greater that one side of the triangle if and only if A lies in the upper arc $\widehat{PQ}$, which is just one third of the whole length of the circumference. Hence we conclude that the probability must be $\frac{1}{3}$.

Second solution: Any member of a system of parallel chords is determined completely by its distance d from the center O. The length of the chord is greater

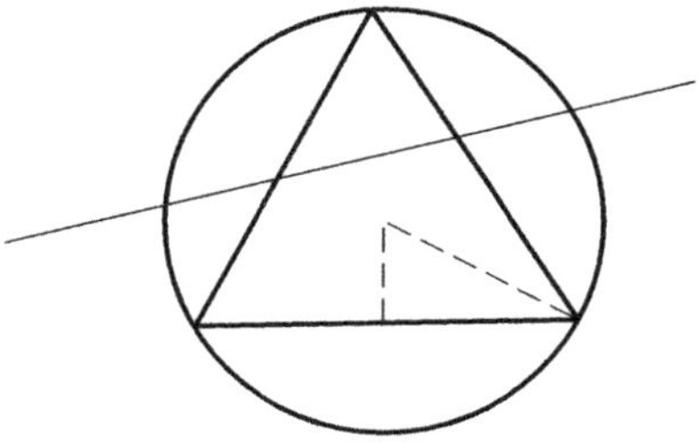

Fig. 2.

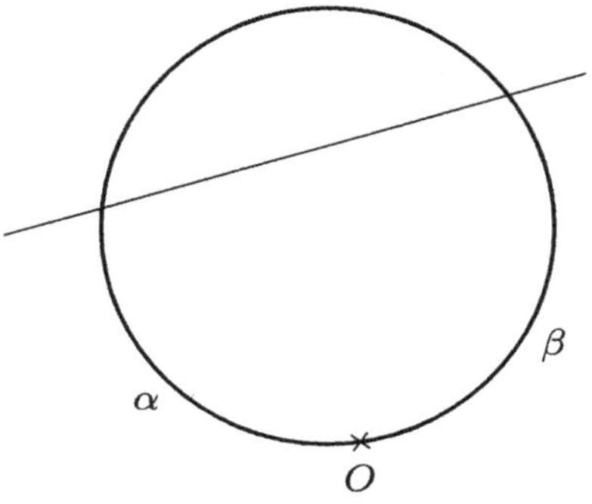

Fig. 3.

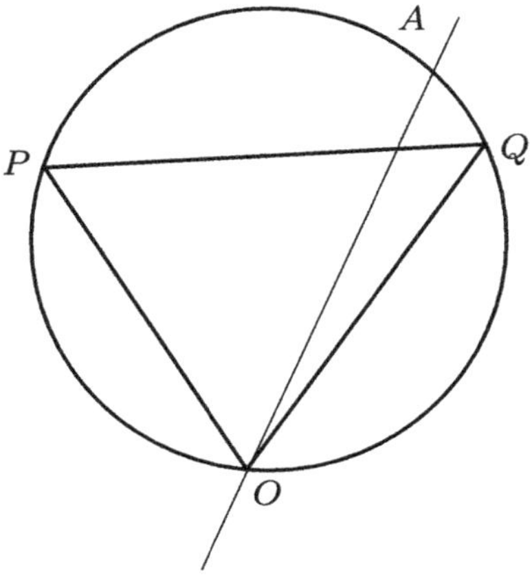

Fig. 4.

than $\sqrt{3}$, (one side of the Δ) if and only if $d < \frac{1}{2}$, since $d < \frac{1}{2}$ is equivalent to $\theta < \frac{\pi}{6}$, i.e. $\cos\theta > \frac{\sqrt{3}}{2}$.

Hence for this system of parallel chords, only one half of them can satisfy the requirement, but this fact holds true for arbitrary system of parallel chords, thus we conclude that the probability must be $\frac{1}{2}$. Ambiguously enough, we have another:

Third solution: A chord is determined completely by its midpoint and it is obvious that a chord is longer than one side of the triangle if and only if the midpoint

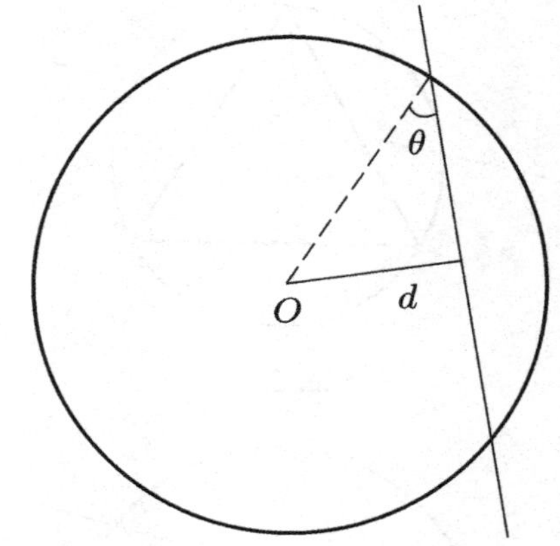

Fig. 5.

lies inside the circle having O as center and $\frac{1}{2}$ as radius. Therefore the probability must be equal to the ratio of the areas of the two concentric circles having $\frac{1}{2}$ and 1 as its radius respectively. This value is $\frac{1}{4}$.

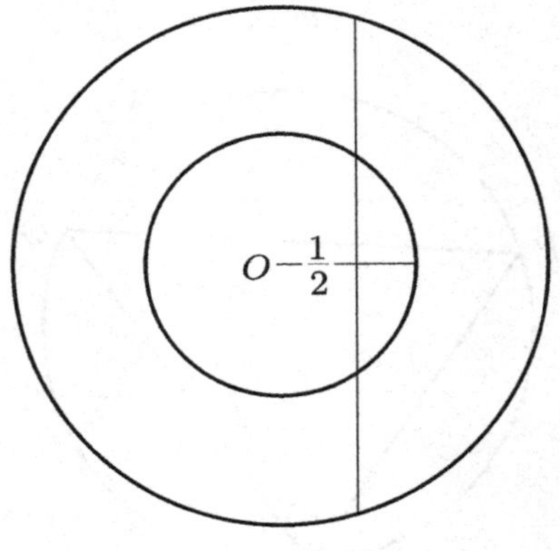

Fig. 6.

What is the right answer? Mr. Bertrand was quite troubled in answering this question. In order to overcome the apparent obstructions, let us consider in this way. A dice having six faces, at every time of throwing it, only one face turns upward, hence we know that the probability of the appearance of a certain face is just $\frac{1}{6}$. In the above case, we are given the set of all lines cutting the unit circle, we want to know the probability that a line belongs to the subset of lines having chord longer than $\sqrt{3}$. If we could assign a measure to the set of all lines intersecting with the unit circle and also assign a measure to the above mentioned subset (just as we have done for a dice, we assign the measure 6 to the set of all possible candidates, and assign the measure 1 to the particular face we hope it to appear), then we could obtain the desired probability as the ratio of these two measures. Hence the problem turns into the one how to determine (or introduce) a measure, for a set of lines on a plane. The measure of a set of points can naturally be defined to be its area. But what is the case for a set of lines?

Measure of a set of lines in the plane:

Let us proceed on this task in a few steps.

i) Let us coordinatize the line first. From O we draw the perpendicular straight line, let p be the distance and θ be its angle with the x-axis, then H has rectangular coordinate $(p\cos\theta,\ p\sin\theta)$. Thus the equation of the straight line is

$$\frac{y-p\sin\theta}{x-p\cos\theta}=\tan\left(\frac{\pi}{2}+\theta\right)=-\cot\theta=\frac{-\cos\theta}{\sin\theta}$$

that is $x\cos\theta+y\sin\theta=p$, which will be called the Hesse normal form of the equation of a straight line.

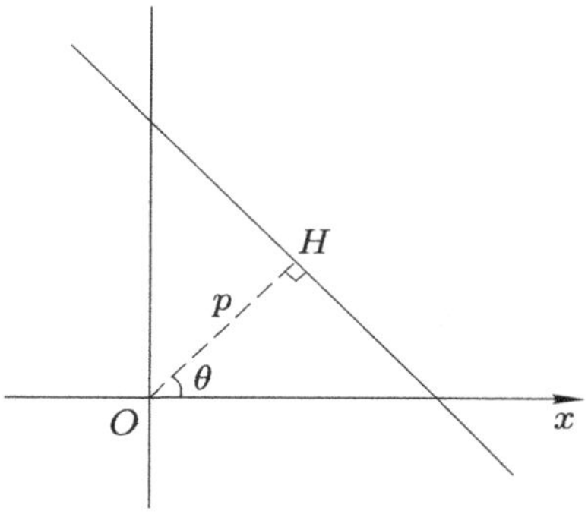

Fig. 7.

Every line is determined completed by (p,θ), thus (p,θ) may be considered as the coordinates of the line. Even at this simple place we can illustrate a deep geometrical concept, which is usually overlooked. We must be very careful to distinguish the delicate and fundamental difference between the line itself and its coordinates (p,θ).

If we said that "$\cdots$ therefore we may regard (p,θ) as the coordinates of the lines, with p, θ ranging over $0\le p<+\infty$, $0\le\theta<2\pi$ respectively", then we would lose an essential concept concerning the orientation. The promising description should be as follows:

We regard p as the distance from the origin, thus it is a real number. The set of all reals will be denoted by R. However, θ is an angle between 0 to 2π, hence it may be considered as a point on the circumference of the unit circle S^1, (we see that the periodicity naturally comes in by this consideration).

Therefore, $(p,\theta)\in R\times S^1$, and we may consider the mapping. $R\times S^1\xrightarrow{\psi}X=$ (the space of all lines on the plane), which carries (p,θ) to the line having the equation

$$x\cos\theta + y\sin\theta = p.$$

It can easily be checked that $\psi(p,\theta) = \psi(p',\theta')$ if and only if

$$p = p', \quad \theta = \theta' \quad \text{or} \quad p = -p', \quad \theta = \theta' + \pi.$$

$R \times S^1$ is the coordinates space of X, hence it is a covering space of X, in fact, it is a two-fold covering space of X.

An element of X is called an unoriented line, while an element of $R \times S^1$ is called an oriented line. A straight line can have two orientations, this just corresponds to the fact that these are two preimages of a line of X in the covering space $R \times S^1$. Hence assigning an orientation to a line in X is just equivalent to taking a coordinate of the line in the covering space.

There are similar treatments for the higher dimensional cases, for example the set of planes in the 3-dim space. The conception of covering space is very important in many fields of mathematics. When looked locally, the sheaves of the covering space have not relations, but when considered globally, these sheaves are connected in some way.

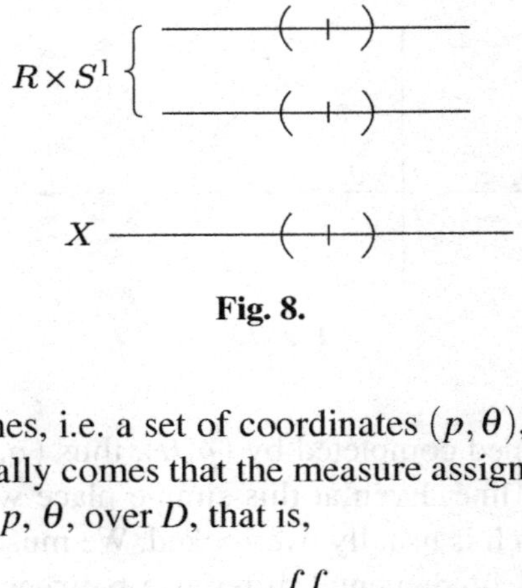

Fig. 8.

ii) Given a set of lines, i.e. a set of coordinates (p,θ), let us denote this given set by D, hence it naturally comes that the measure assigned to this set must be an integral with respect to p, θ, over D, that is,

$$\text{measure} = \iint_D dp\, d\theta.$$

But we may consider in a more general way and take

$$\text{measure} = \iint_D F(p,\theta) dp\, d\theta,$$

since this measure depends on p, θ, where $F(p,\theta)$ is a continuous function of p, θ.

In the case of a set of points, the measure is defined by

$$\text{measure} = \iint_D f(x,y) dx\, dy.$$

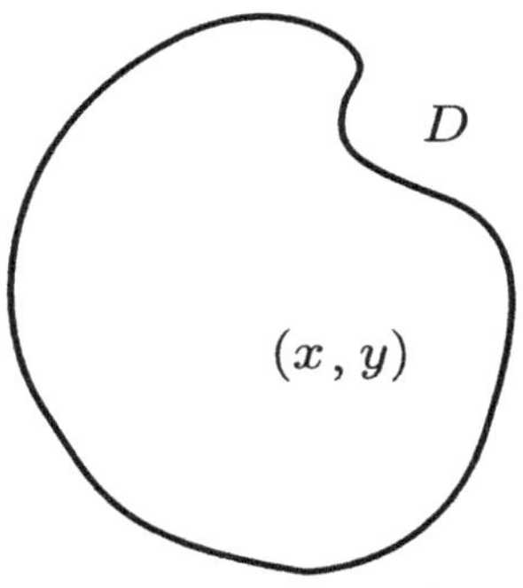

Fig. 9.

For simplicity, we assume $f(x,y)$ to be positive continuous function. It is natural to require that such a measure is invariant under the group of motions in the plane, since the area is invariant under rotations or translations. It can easily be checked that (cf. Santalo p. 6, or the method which will be used later), in order that the measure has this invariant property it is necessary, and sufficient to put $f(x,y)$ to be a constant factor. Thus without loss of generality, the measure of a set D of points in the plane is defined by $\iint_D dx\,dy$.

We shall in that same fashion require that the measure of a set of lines also enjoys this property of invariance under the group of motions.

The general formulas for the motion are

$$\begin{cases} x = a + x^* \cos\alpha - y^* \sin\alpha, \\ y = b + x^* \sin\alpha + y^* \cos\alpha, \end{cases} \tag{1}$$

where a, b, α are constants. The line having equation $x\cos\theta + y\sin\theta = p$ becomes a line with equation

$$(a + x^* \cos\alpha - y^* \sin\alpha)\cos\theta + (b + x^* \sin\alpha + y^* \cos\alpha)\sin\theta - p = 0$$

under the above motion. That is,

$$x^* \cos(\theta - \alpha) + y^* \sin(\theta - \alpha) = p - a\cos\theta - b\sin\theta$$

or

$$x^* \cos\theta^* + y^* \sin\theta^* = p^*.$$

Hence we see that the motion (1) transforms the line (p,θ) to the line (p^*,θ^*) which is related to (p,θ) by

$$\begin{cases} p^* = p - a\cos\theta - b\sin\theta, \\ \theta^* = \theta - \alpha. \end{cases}$$

Since the Jacobian

$$\frac{\partial(p^*,\theta^*)}{\partial(p,\theta)} = \begin{vmatrix} \frac{\partial p^*}{\partial p} & \frac{\partial p^*}{\partial \theta} \\ \frac{\partial \theta^*}{\partial p} & \frac{\partial \theta^*}{\partial \theta} \end{vmatrix} = \begin{vmatrix} 1 & * \\ 0 & 1 \end{vmatrix} = 1.$$

We conclude that $\iint_D d\, pd\theta = \iint_{D^*} dp^*\, d\theta^*$, that is, if we take $F(p,\theta) = 1$ and define the measure to be equal to $\iint_D dpd\theta$, then it is invariant under motions. Let us formulate it formally:

Theorem 3.1.1. *The measure of a set of lines given by $\iint_D dp\, d\theta$ is invariant under motion. Moreover, up to a constant factor, this is the only measure invariant under motion.*

Let us prove the second part of the theorem. Let $\iint\limits_D F(p,\theta)dpd\theta$ be another invariant measure over any domain D, we would like to know what can the values of $F(p,\theta)$ assume?

Since $\frac{\partial(p^*,\theta^*)}{\partial(p,\theta)} = 1$. We have by the assumption and by the rule of change of variables

$$\iint\limits_D F(p,\theta)dp\, d\theta = \iint\limits_{D^*} F(p^*,\theta^*)dp^*\, d\theta^* = \iint\limits_D F(p^*,\theta^*)dp\, d\theta$$

i.e.

$$\iint\limits_D \{F(p,\theta) - F(p^*,\theta^*)\}dp\, d\theta = 0$$

for all domains D.

From the continuity of $F(p,\theta)$, We conclude that $F(p,\theta) = F(p^*,\theta^*)$. But the group of motions is transitive in the space X,i.e. for arbitrary (p,θ), (p^*,θ^*) there exists a motion which transforms (p,θ) to (p^*,θ^*). Therefore $F(p,\theta) = $ constant. This proves the theorem.

This constant factor is not important at all, because it only represents the choice of the unit of the measurement in our space X. Later we shall always take the measure to be $\iint_D dp\, d\theta$ and call $dp\, d\theta$ the density of our set of lines.

Remark 3.1.2. *For simplicity we shall use a simple integral sign to denote all, integrations, if this does not lead to confusion.*

Remark 3.1.3. *The density has been defined as the differential form $dp\, d\theta$, hence if we want to express it by means of other variables p^*, θ^* given by $p = p(p^*,\theta^*)$, $\theta = \theta(p^*,\theta^*)$. We must have*

$$dp\, d\theta = \frac{\partial(p,\theta)}{\partial(p^*,\theta^*)} dp^* d\theta^*. \qquad (2)$$

Thus, instead of the ordinary multiplication of the differentials

$$dp = p_{p^*}dp^* + p_{\theta^*}d\theta^*, \; d\theta = \theta_{p^*}dp^* + \theta_{\theta^*}d\theta^* \quad where \; p_{p^*} = \frac{\partial p}{\partial p^*} \; etc.,$$

we must apply, the rule (2). In this case we shall say that the multiplication of dp by $d\theta$ is exterior. To avoid the confusion, we shall write $dp \wedge d\theta$ (called dp wedge $d\theta$) to denote this exterior multiplication. Its properties can be characterized by the fact that it is antisymmetrics, i.e. $dp \wedge d\theta = -d\theta \wedge dp$. As a consequence we have $dp \wedge dp = 0$.

For example, in the above situation.

$$\begin{aligned} dp \wedge d\theta &= \left(\frac{\partial p}{\partial p^*}dp^* + \frac{\partial p}{\partial \theta^*}d\theta^*\right) \wedge \left(\frac{\partial \theta}{\partial p^*}dp^* + \frac{\partial \theta}{\partial \theta^*}d\theta^*\right) \\ &= \frac{\partial p}{\partial p^*}\frac{\partial \theta}{\partial \theta^*}dp^* \wedge d\theta^* + \frac{\partial p}{\partial \theta^*}\frac{\partial \theta}{\partial p^*}d\theta^* \wedge dp^* \\ &= \frac{\partial(p,\theta)}{\partial(p^*,\theta^*)}dp^* \wedge d\theta^*. \end{aligned}$$

Thus by using exterior multiplication, we need not trouble ourselves with the change of variables, since the Jacobian naturally associated with this multiplication. Hence it can simplify a great deal of our treatments. Moreover since we will exclude negative densities, we will always take the densities in absolute values.

Having equipped with this knowledge of measure, we are well prepared to overcome the Bertrand Paradox.

In the first solution, the lines are determined by α, β which in turns can be expressed as functions of p, θ. From this we can calculate its density and obtain

$$d\alpha \wedge d\beta = \frac{2}{\sqrt{1-p^2}}dp \wedge d\theta. \tag{3}$$

As for the second solution, the density is just $dp \wedge d\theta$.

For the third solution, it is area that plays the role, that is, its density is expressed by

$$pdp \wedge d\theta. \tag{4}$$

We see, that (3) and (4) are not invariant under the Erouo of motions, thus only the second solution is admissible. And the right answer is that the probability is equal to $\frac{1}{2}$.

Usually we denote the density by the small German letter $\mathfrak{y} = dp \wedge d\theta$, and so measure $= \int_D \mathfrak{y}$.

Formulas of Crofton:

Let C be a fixed smooth curve, that is, a curve with tangent at each point. We assume that, C has a finite length L and write its equation as $x = x(s)$, $y = y(s)$ where the parameter s is the arc length, therefore $(x'(s), y'(s)) \neq (0,0)$ everywhere. The straight line (p,θ) intersects with C at the point $P(x,y)$. As line coordinates of lines meeting C we can have (p,θ) and (s,φ), where φ denotes the angle between

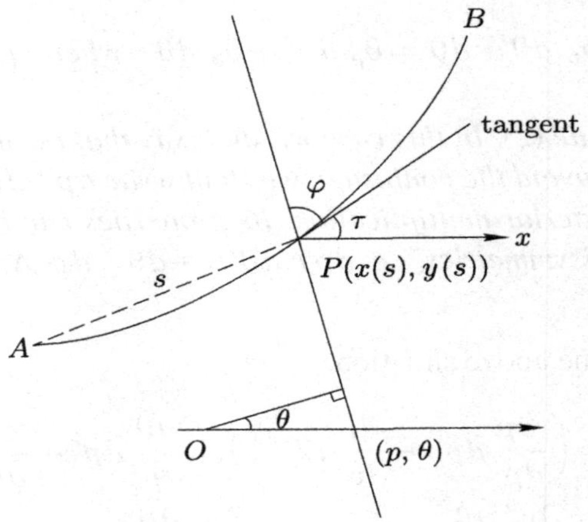

Fig. 10.

the tangent and the line considered, τ denotes the angle inclination of the tangent at P. Then

$$\cos\tau = \frac{dx}{ds},\ \sin\tau = \frac{dy}{ds} \quad \text{and} \quad \theta = \tau(s) + \varphi \pm \frac{\pi}{2} \tag{5}$$

is the inclination of the perpendicular line (to the line (p,θ)) with respect to the x-axis.

The equation of (p,θ) is

$$p = x(s)\cos\theta + y(s)\sin\theta. \tag{6}$$

From (5), (6) we obtain the relations associating the two coordinates. Now let us calculate the density:

$$\begin{aligned} dp\wedge d\theta &= \{x'ds\cos\theta + y'ds\sin\theta\}\wedge d\theta \\ &= \{x'\cos\theta + y'\sin\theta\}ds\wedge d\theta \\ &= \{\cos\tau\cos\theta + \sin\tau\sin\theta\}ds\wedge d\theta \\ &= \cos(\tau-\theta)ds\wedge d\theta \\ &= \cos(\theta-\tau)ds\wedge d\theta \\ &= \cos\left(\varphi \pm \frac{\pi}{2}\right) ds\wedge d\varphi. \end{aligned}$$

But $d\theta = \tau' ds + d\varphi$, hence

$$\eta = dp\wedge d\theta = \left|\cos\left(\varphi\pm\frac{\pi}{2}\right)\right| ds\wedge d\varphi = |\sin\varphi| ds\wedge d\varphi \tag{7}$$

where we have written $|\sin\theta|$ because we consider all densities to be positive. We Integrate both sides of (7) over all straight-lines which intersect C. On the right hand side, by fixing s, (i.e. the point P) first we will have:

$$\int_0^L ds \int_0^\pi |\sin\varphi| d\varphi = 2L.$$

On the left hand side, each straight line (p, θ) has been counted as many times as it has intersection points with C. Calling this number n, we obtain

$$\int_n \mathfrak{y} = 2L. \tag{8}$$

This is called the formula of (Cauchy-) Crofton.

If a line does not intersect the curve C, then the n associated to this line is zero, hence our notation is rather convenient, and (8) can be regarded as integrated over the set of all straight lines on the plane.

Remark 3.1.4. *In fact, this formula holds for any rectifiable curve. A rectifiable curve is a continuous curve, such that we can choose a parameter t, and when t runs from* 0 *to* 1, $\delta(t)$ *just runs over the curve one time, and after suitable definition, the curve has finite arc length. In this case, n may become infinite for some lines, we have to prove that the measure of such lines is zero, hence they can be neglected, and the formula remains meaningful. The detailed proof for this more general case can be found in Blaschke: Vorlesungen über Integralgeometrie. p. 46.*

Moreover, Favard, in C. R. Academie de France, 1932, takes the left hand side of (8) as the definition of length of a continuum of points and proves that it coincides with the ordinary length in the case of rectifiable curves. This method of approach is of interest and importance. Because of this same approach we may define the area of a smooth surface. And if the surface is not smooth, we can use it at least as a reference of considering how to define the area, or whether the area of such surface exists or not. This problem is rather complicate.

Corollary 3.1.5. *If C is a convex curve, we have* $n = 2$*, for all lines which intersect C, except for the tangents* ($n = 1$, $\varphi = 0$, *thus* $ds \wedge d\varphi = 0$)*, which are of measure zero. Consequently we have,*

$$\int_{\ell \cap C \neq \emptyset} \mathfrak{y} = L. \tag{9}$$

That is: the measure of the set of straight lines which intersect a convex curve, is equal to its length.

Corollary 3.1.6. *If C is a segment of length L, we can construct a sequence of convex curve approaching to this segment. Hence*

$$\int_{\ell \cap C \neq \emptyset} \mathfrak{y} = 2L.$$

Let us suppose that a curve C of length L can be contained inside a convex C_1 *of length* L_1*. If we consider all straight lines that intersect* C_1*, the mean value of the number of the intersection points of these straight lines with C will be*

$$\overline{n} = \frac{\int_n \mathfrak{y}}{\int \mathfrak{y}} = \frac{2L}{L_1}. \tag{10}$$

Since there must be set of lines of positive measure the number of whose intersections with C is not less than $\overline{n}$, we have shown:

Corollary 3.1.7. *If a curve of length L can be contained inside a convex curve of length L_1, then there exists a set of straight lines of positive measures, each of which intersects C in $n \geqslant \frac{2L}{L_1}$ points.*

The longer the length of C, the greater the value n. This fact coincides with our common sense.

3.2 Lecture II

The Cauchy-Crofton formula $\int_n \mathfrak{y} = 2L$ has several important corollaries. These corollaries are obtained by considering the particular case when C is a convex (closed) curve. The discussions concerning convex domains are very broadly found in mathematics, especially, e.g. in finite dimensional space or geometric number theory, Minkowski has done a great deal in this respect. Moreover, when we generalize the convex properties to Hilbert or Banach spaces, they become very important conceptions in analysis. But what we shall be concerned here of the convex properties are only those of the simplest.

We have discussed Cors. 3.1.5 and 3.1.6 in the last lecture, now we begin from the following consideration. Consider a convex curve C, and consider any support line (p, θ) which is tangent to C, clearly p is a function of θ, we call this function $p(\theta)$ the support function. Therefore a convex curve (or the convex domain it contains) is determined completely by the set of support functions. (Note that for a closed curve which is not convex, we can still consider its support line by selecting the first tangent which intersect the curve only at the tangent point.) By Crofton's formula $\iint dp\, d\theta = L$.

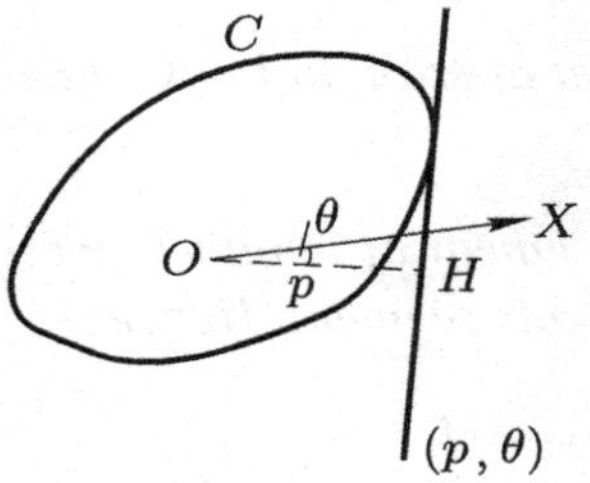

Fig. 11.

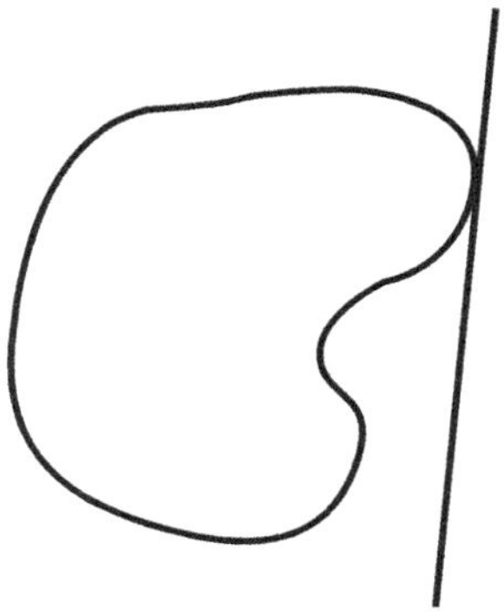

Fig. 12.

Fixing θ, by Fubini's theorem we have for the left hand side $\int_{-\pi}^{\pi}(\int_0^H dp)d\theta = \int_{-\pi}^{\pi} p(\theta)d\theta$ since $\int_0^H dp = [p]_0^H = p(\theta) - 0 = p(\theta)$. Thus we obtain the result:
Corollary 3.2.2 a) $\int_{-\pi}^{\pi} p(\theta)d\theta = L$, *which is called the Cauchy formula.*

Definition 3.2.1. Let C be a convex curve, it is called a curve of constant breadth if $p(\theta) + p(\theta + \pi) = b =$ constant. Where $p(\theta)$is the support function, this means that all pairs of parallel tangents (supporting lines) have a constant distance b, for example, a circle has this property.

There is a question which was proposed very early: Except the circle, does there exist any other curve of constant breadth? The answer is affirmative. There are convex curves, other than the circle which are of constant breadth, e.g. the Reuleux polygons, which can be found in Rlaschke: Vorlesungen über differential geometrie, Vol. 2.

Since $p(\theta)$ has 2π as its period, we have also from Cauchy formula $\int_{-\pi}^{\pi} p(\theta + \pi)d\theta = L$, since it is integrated over a period of 2π. Combining the two, we have $\int_{-\pi}^{\pi}\{p(\theta) + p(\theta + \pi)\}d\theta = 2L$. From which we have:

Corollary 3.2.2. b). *If C is a curve of constant breadth b, then*

$$2\pi b = 2L, \text{ that is, } \ L = \pi b.$$

This is an extension of the formula for the circle.

Corollary 3.2.3. *Suppose that an arc C is contained in a convex domain whose boundary is C_1. C and C_1 are of lengths L and L_1 respectively. The Crofton's formulas are $\int_{\forall \ell^*} n\mathfrak{y} = 2L$ and $\int_{\ell \cap C_1 \neq \emptyset} \mathfrak{y} = L_1$ for the two curves respectively. Since $C \subset C_1$, any line ℓ intersecting with C must intersect with C_1 also. If we denote by X, the set of all lines; A, (B resp.) the set of all lines intersecting C_1 (C respectively), then we have $B \subset A \subset X$.*

If a line lies in X, B, then the number n for this line is equal to zero, hence we have $\int_X n\mathfrak{y} = \int_A n\mathfrak{y} = \int_B n\mathfrak{y} = 2L$ for the curve C.

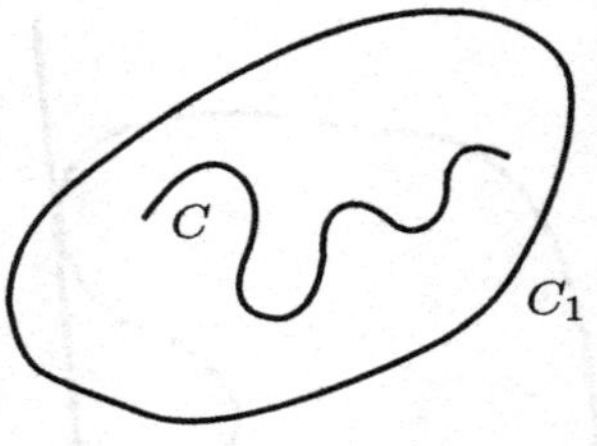

Fig. 13.

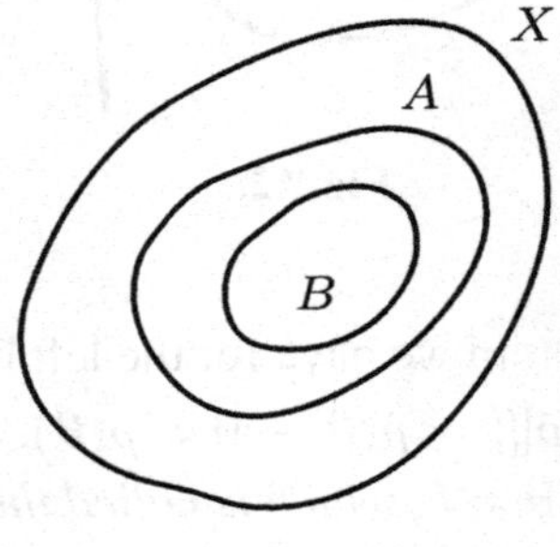

Fig. 14.

Dividing by $2L$ and L_1 respectively in

$$\int_A n\mathfrak{y} = 2L \qquad \text{and} \qquad \int_A \mathfrak{y} = L_1,$$

we have

$$\int_A \frac{n\mathfrak{y}}{2L} = 1 = \int_A \frac{\mathfrak{y}}{L}$$

or

$$\int_A \left(\frac{n\mathfrak{y}}{2L} - \frac{\mathfrak{y}}{L_1}\right) = 0$$

or

$$\int_A \left(n - \frac{2L}{L_1}\right)\mathfrak{y} = 0.$$

Thus $n - \frac{2L}{L_1} \geq 0$ for a set of lines of positive measure, since if for all sets of positive measure $n - \frac{2L}{L_1} \geq 0$ does not hold, then $n - \frac{2L}{L_1} < 0$ must hold. Especially for A, $n - \frac{2L}{L_1} < 0$ therefore $\int_A (n - \frac{2L}{L_1})\mathfrak{y} \neq 0$, a contradiction. This inequality enables us to assert that there exists a set of lines of positive measure such that each of which intersects with C at least in $[\frac{2L}{L_1}]$ points, where $[\frac{2L}{L_1}]$ denotes the least integer greater than or equal to the number $\frac{2L}{L_1}$. This means that the longer the line C, the more we have intersecting points with it.

Of course in the argument above, it is obvious that we have also $n - \frac{2L}{L_1} \leq 0$ for a set of lines of positive measure, but this has no important geometric meaning.

Corollary 3.2.4. *Solution of Buffon's needle problem.*

The plane is equipped with a set of parallel lines with constant distance a, and we are given a needle of length $L < a$. Now let us put the needle upon a disk with diameter a. Instead of throwing the needle upon the plane, we throw the disk. Except for a set of measure zero, the disk always intersects with one and only one line, hence our problem is completely equivalent to the consideration of the comparison of the measures of lines intersecting the disk and that of intersecting the line segment of length L which is contained in the disk. The measure of the set of lines intersecting the disk is equal to the length of its boundary πa; whereas the measure of those lines intersecting the line segment is $2L$. This means that among the set of lines of measure πa, only a part of it (of measure $2L$) intersects the needle. And therefore the probability is $\frac{2L}{\pi a}$.

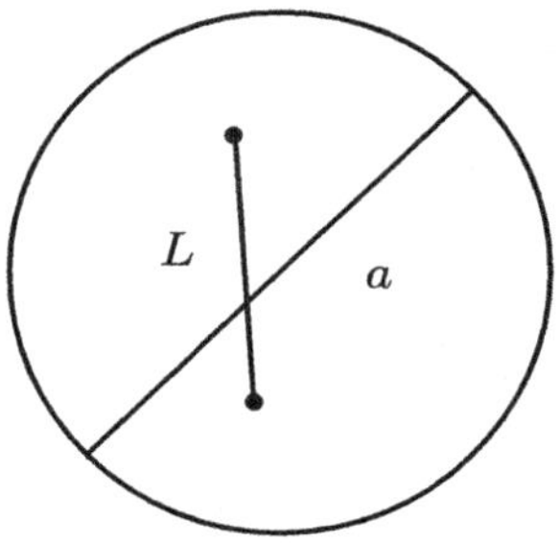

Fig. 15.

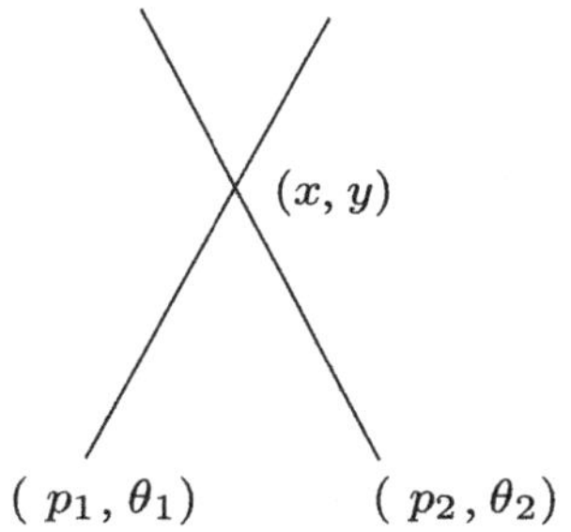

Fig. 16.

Another formula of Crofton.

Let us consider two lines (p_1, θ_1), (p_2, θ_2), if (x, y) is their intersection point. The coordinates of this pair may be written in many ways, for example,

$$(p_1, p_2, \theta_1, \theta_2) \qquad \text{or} \qquad (x, y, \theta_1, \theta_2),$$

since each of them determines completely the lines pair.

A transformation can be found existing between these two coordinates, it is worthwhile to notice and in fact it is very interesting that in integral geometry we can frequently derive important formulas from these coordinate transformation formulas. We shall become more and more conscious of this fact, since we shall encounter many examples of this kind later.

If $\mathfrak{y}_1, \mathfrak{y}_2$ denote the densities of $(p_1, \theta_1), (p_2, \theta_2)$, respectively, let us define the density of the pair of lines to be

$$\text{density } = \mathfrak{y}_1\mathfrak{y}_2 = dp_1 \wedge d\theta_1 \wedge dp_2 \wedge d\theta_2.$$

This definition is well-defined, because it is invariant under motion.

It should be noticed that by integrating we always take the absolute value of the differential form (the density), this is used in measure theory also. But in the cohomology theory we encounter also the problem of integrating a differential form, in this case we assign the orientations (plus or minus, symbol) to differential forms and integrate with the consideration of sign. These form the two branches in the integration theory of differential forms, both are important and should not be confused, sometimes their results coincide, but often they do differ from each other.

Many complicated conceptions frequently appear in the simple examples we are treating of. Hence by these simple examples we can visualize these complicated ideas more clearly and easily, however, in some cases, it is more convenient and easy to treat these abstract complicated ideas directly, because in the simple examples only the "locus" of the conceptions appears, thus in order to acquaint with them completely and thoroughly, we have to do directly.

In my opinion, I think it is important for a mathematician to train himself to be able both to master the general abstract conceptions and to have a wide knowledge of special examples. Only through that, can he reach a level, which is necessary for him to become a good mathematician.

OK. Let's return to our density $\mathfrak{y}_1\mathfrak{y}_2$ for a pair of lines. Since (x,y) is the intersection, it satisfies both the two lines of the pair, hence $x\cos\theta_i + y\sin\theta_i = p_i$ holds for both $i = 1, 2$. By differentiating

$$\cos\theta_i dx + \sin\theta_i dy + (\cdots)d\theta_i = dp_i$$

thus

$$\begin{aligned} dp_1 \wedge dp_2 \wedge d\theta_1 \wedge d\theta_2 &= (dx\cos\theta_1 + dy\sin\theta_1) \wedge (dx\cos\theta_2 + dy\sin\theta_2) \wedge d\theta_1 \wedge d\theta_2 \\ &= \sin(\theta_2 - \theta_1) dx \wedge dy \wedge d\theta_1 \wedge d\theta_2. \end{aligned}$$

Consider a convex domain K, with boundary curve γ. Consider lines ℓ_1, ℓ_2 such that $\ell_i \cap K \neq \emptyset$, and $A =$ area of K, $L =$ length of γ. Since ℓ_1 and ℓ_2 are independent from, each other, we have

$$I \equiv \int_{\ell_i \cap K \neq \emptyset} \mathfrak{y}_1\mathfrak{y}_2 = \int_{\ell_1 \cap K \neq \emptyset} \mathfrak{y}_1 \int_{\ell_2 \cap K \neq \emptyset} \mathfrak{y}_2 = L^2$$

on the other hand, $I = \int_{\ell_i \cap K \neq \emptyset} |\sin(\theta_1 - \theta_2)| dx\, dy\, d\theta_1\, d\theta_2 \equiv I_{int} + I_{ext}$, where $I_{int} = \int_{\ell_1 \cap \ell_2 \cap K \neq \emptyset} \mathfrak{y}_1 \mathfrak{y}_2$, $I_{ext} = \int_{\ell_1 \cap \ell_2 \cap K = \emptyset, \ell_i \cap K \neq \emptyset} \mathfrak{y}_1 \mathfrak{y}_2$.

If the intersection point (x, y) lies outside of K, the two tangents issuing from (x, y) have an angle ψ. Clearly, in order that $\ell_i \cap K \neq \emptyset$, (taking one tangent as reference axis) the inclination angles of ℓ_1, ℓ_2 are limited to $0 \leq \theta_i \leq \psi$, while if $(x, y) \in K$, (taking any line through (x, y) as reference) the inclination angles of ℓ_1, ℓ_2 can assume any value in $0 \leq \theta_i \leq \pi$.

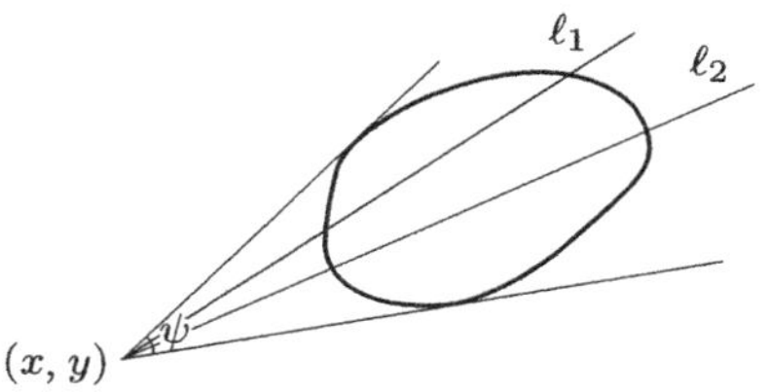

Fig. 17.

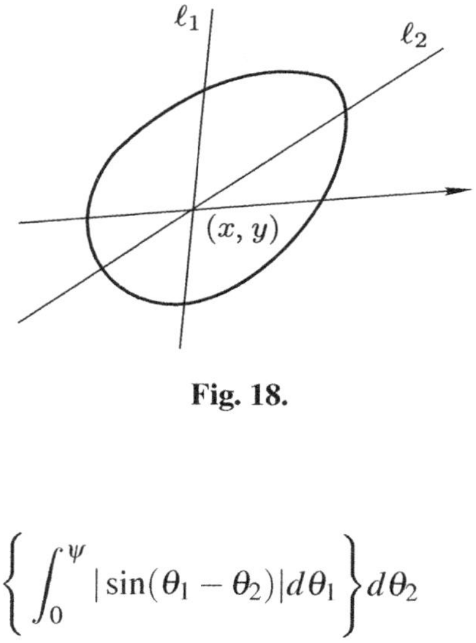

Fig. 18.

Let us calculate

$$
\begin{aligned}
&\int_0^{\psi} \left\{ \int_0^{\psi} |\sin(\theta_1 - \theta_2)| d\theta_1 \right\} d\theta_2 \\
=& \int_0^{\psi} \left\{ (\cos(\theta_1 - \theta_2)]_0^{\theta_2} - \cos(\theta_1 - \theta_2)]_{\theta_2}^{\psi} \right\} d\theta_2 \\
=& \int_0^{\psi} \{(1 - \cos\theta_2) - \cos(\psi - \theta_2) + 1\} d\theta_2 \\
=& 2\theta_2|_0^{\psi} - \sin\theta_2|_0^{\psi} + \sin(\psi - \theta_2)|_0^{\psi} \\
=& 2\psi - 2\sin\psi \\
=& 2(\psi - \sin\psi).
\end{aligned}
$$

Therefore $I_{int} = \int_K 2(\pi - \sin\pi) dx\, dy = 2\pi A$.

Since I_{ext} is integrated over K^c, the complement of K, it is an improper integral, nevertheless its convergence has already been assured from the fact that both I and I_{int} exist, and $I_{ext} = I - I_{int}$. Therefore $I_{ext} = 2\int_{(x,y)\notin K, \ell_i \cap K \neq \emptyset}(\psi - \sin\psi)dxdy = L^2 - 2\pi A$ and we have proved the formula $\int_{(x,y)\notin K, \ell_i \cap K \neq \emptyset}(\psi - \sin\psi)dxdy = \frac{1}{2}L^2 - \pi A$, which is also called the Crofton's formula.

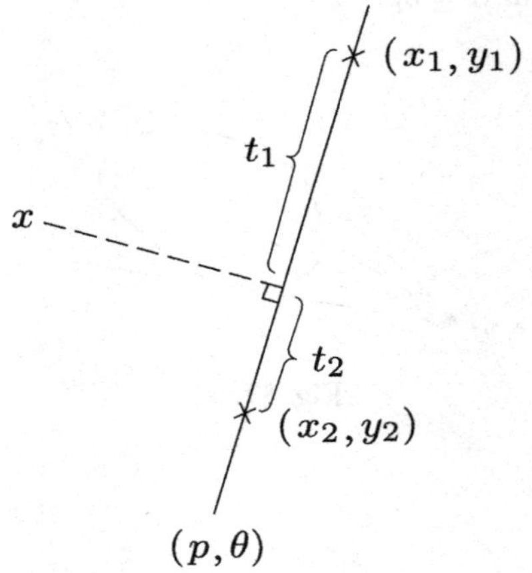

Fig. 19.

Similarly for a pair of points (x_1,y_1), (x_2,y_2), it may be determined by four coordinates (x_1,y_1,x_2,y_2). It may also be determined, by the coordinates (p,θ) of the straight line determined by (x_1,y_1), (x_2,y_2), together with the distances t_1, t_2 from the two points, to the foot of the perpendicular from 0 to (p,θ). We want to express the product, $dx_1 \wedge dy_1 \wedge dx_2 \wedge dy_2$, which is defined to be the density of this points pair, by means of the coordinates p, θ, t_1, t_2. We have

$$x_i = p\cos\theta - t_i \sin\theta, \qquad y_i = p\sin\theta + t_i\cos\theta \ (i = 1,2).$$

Hence

$$dx_i = \cos\theta dp - \sin\theta dt_i - (p\sin\theta + t_i\cos\theta)d\theta,$$
$$dy_i = \sin\theta dp + \cos\theta dt_i + (p\cos\theta - t_i\sin\theta)d\theta,$$

thus $dx_i \wedge dy_i = p[dp,d\theta] + [dp,dt_i] - t_i[d\theta,dt_i]$.

Taking $i = 1,2$, and multiplying again, we see that

$$\begin{aligned} dx_1 \wedge dy_1 \wedge dx_2 \wedge dy_2 &= (t_2 - t_1)dp \wedge d\theta \wedge dt_1 \wedge dt_2 \\ &= (t_2 - t_1) \wedge dt_1 \wedge dt_2. \end{aligned}$$

Because all densities are assumed to be positive, we take $|t_2 - t_1|\eta dt_1 dt_2$ as the density of a pair of points with coordinates (p,θ,t_1,t_2).

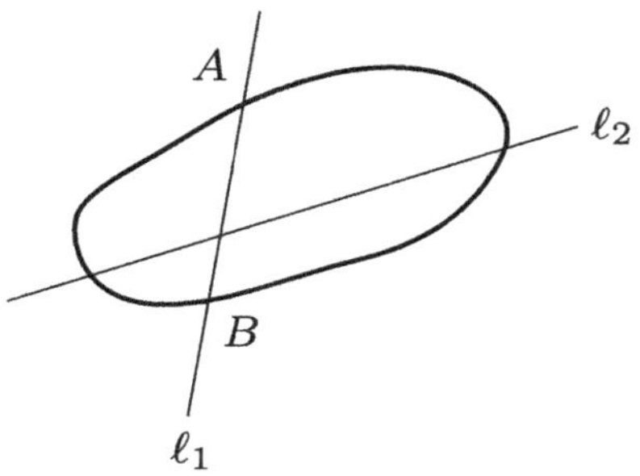

Fig. 20.

As has been shown $I_{int} = 2\pi A$, we shall translate $I_{int} = \int_{\ell_1 \cap \ell_2 \cap K \neq \emptyset} \mathfrak{y}_1 \mathfrak{y}_2$ into another form. Fix ℓ_1 first, and consider all ℓ_2 which intersects with ℓ_1 in K, i.e. intersects with ℓ_1 in the chord AB. By Crofton's formula, the set of lines ℓ_2 considered above has measure $2s$, where s denotes the length of the chord AB. We have therefore

$$2\pi A = I_{int} = \int_{\ell_1 \cap K \neq \emptyset} \mathfrak{y}_1 \int_{\ell_2 \cap (\ell_1 \cap K) \neq \emptyset} \mathfrak{y}_2 = \int_{\ell_1 \cap K \neq \emptyset} \mathfrak{y}_1 2s = 2 \int_{\ell_1 \cap K \neq \emptyset} s\mathfrak{y}_1 .$$

Let ℓ_1 vary, and omitting the subscript, we obtain the formula

$$\int_{\ell \cap K \neq \emptyset} s\mathfrak{y} = \pi A.$$

And we already have the Crofton's formula $\int_{\ell \cap K \neq \emptyset} \mathfrak{y} = L$. From these two formulas, we see that we may associate to a convex curve a sequence of numbers $\{\sigma_k\}$ defined by $\sigma_k = \int_{\ell \cap K \neq \emptyset} s^k \mathfrak{y}$, $k = 0, 1, 2, \cdots$.

Let us make, a few calculations of these numbers by using the density for points pair (P_1, P_2), which will be denoted by $dP_1 \wedge dP_2$ (i.e. $dx_1 \wedge dy_1 \wedge dx_2 \wedge dy_2$) for convenience. Besides $\sigma_k = \int_{\ell \cap K \neq \emptyset} s^k \mathfrak{y}$, we shall also consider the integrals $J_k = \int \gamma^k dP_1 dP_2$, where γ is the distance between the two points P_1, P_2, which are supposed to be inside K, the integration being extended over all such pairs of points. Since $dP_1 dP_2 = |t_2 - t_1| \mathfrak{y} dt_1 dt_2$, and note that $\gamma = |t_2 - t_1|$ we get:

$$\begin{aligned} J_k &= \int |t_2 - t_1|^{k+1} \mathfrak{y} dt_1 \cdot dt_2 \\ &= \int \mathfrak{y} dt_1 \left[\int_{t_1}^{b} (t_2 - t_1)^{k+1} dt_2 + \int_{a}^{t_1} (t_1 - t_2)^{k+1} dt_2 \right] \\ &= \frac{1}{k+2} \int \mathfrak{y} \int_{a}^{b} (b - t_1)^{k+2} + (t_1 - a)^{k+2} dt_1 \\ &= \frac{2}{(k+2)(k+3)} \int (b-a)^{k+3} \mathfrak{y}, \end{aligned}$$

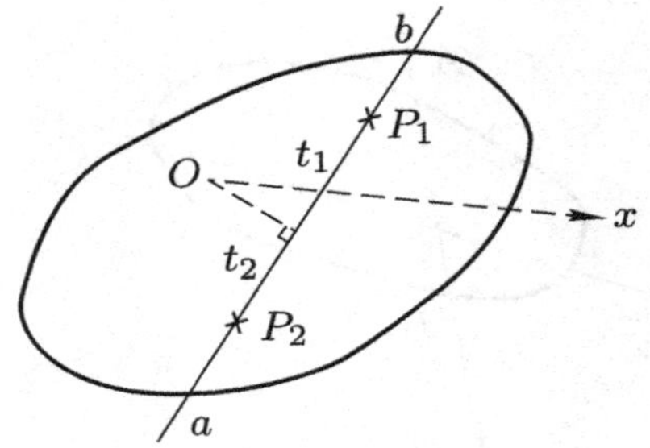

Fig. 21.

where b and a signify the values of t corresponding to the end points of the chord considered, so that $b-a=s$.

We have then $J_k = \frac{2}{(k+2)(k+3)}\sigma_{k+3}$ which holds for $k \geq -1$. This can also be written as the inequality $\sigma_k = \frac{k(k-1)}{2}J_{k-3}$ which holds for $k \geq 2$.

For the cases $k = 0, 1$, we have already obtained

$$\sigma_0 = L, \qquad \sigma_1 = \pi A.$$

For $n = 2$, we have $\sigma_2 = \int s^2\eta = \frac{2\cdot 1}{2}\cdot J_{-1} = \int \frac{dP_1dP_2}{\gamma}$.

For $n = 3$, $\sigma_3 = \int s^3\eta = \frac{3\cdot 2}{2}J_0 = 3\int dP_1dP_2 = 3A^2$. This equality is a remarkable integral formula due to Crofton. Denoting by $\overline{\gamma}$ the mean distance between two points inside K, so that $\overline{\gamma} = \frac{\int \gamma dP_1dP_2}{A^2}$, we have for $n = 4$ the equation $\sigma_4 = \int s^4\eta = 6J_1 = 6\int \gamma dP_1dP_2 = 6\overline{\gamma}A^2$.

Thus we associate to a convex curve a set of positive real numbers which have certain relations with L and A. Conversely we ask: Given a set σ_k' of positive numbers, under what conditions does there exist a convex domain K (i.e. a convex curve ∂K), such that $\sigma_k(K) = \sigma_k'$ and is this convex curve unique? This is a very difficult problem proposed by W. Blaschke, it still remains unsolved. This problem also appears in the moment problem of probability theory. Similar problems often arise in different fields of mathematics, for example, the work of J. Eells of characterizing a differentiable manifold from the considerations of eigenfunctions and eigenvalues of the Laplace equation.

As for the above problem, we have a certain results concerning the necessary conditions. We have known:

$$\sigma_0^2 - 4\sigma_1 \geq 0, \text{i.e. } L^2 - 4\pi A \geq 0.$$

$$15^2\pi^8\sigma_4^2 - 2^{16}\sigma_1^5 \geq 0 \text{ (Blaschke. Math. Zeitschrift. v. 1)},$$

$$16^2\sigma_1^3 - 3^2\pi^4\sigma_2^2 \geq 0 \text{ (Carleman. Math. Zeitschrift. v. 3)}.$$

The equality holds only for the case of a circle. These inequalities are proved using the method of symmetrization. From these results I make a conjecture that there is an inequality relating σ_1 and σ_{2m} with the form $\sigma_1^{\text{power}} - c\sigma_{2m}^{\text{power}} \geq 0$ for m odd, and $c\sigma_{2m}^{\text{power}} - \sigma_1^{\text{power}} \geq 0$ for m even.

The first one $L^2 - 4\pi A \geq 0$ is called the isoperimetric inequality, since the equality sign holds iff K is a circle, we know that with constant length of boundary, the circle has the greatest area. Or exactly the same, with constant circle has the greatest area. Or exactly the same, with constant area, the circle has the shortest boundary. Later we shall prove this formula by expanding $L^2 - 4\pi A$ into a sum of non-negative numbers, each of which has a geometric meaning. This method is due to Santalo and perhaps may be considered as the most important contribution given by him to the study of integral geometry. There are two periods of the recent developments of integral geometry:

Blaschke – Santalo: 1934–
Gelfand – Helgason: 1961–

The work of Gelfand and Helgason concentrated in the applications of integral geometry to the representation theory of non-compact Lie group and vice versa. As a matter of fact, the representation theory of Lie groups considered in the non-compact Hilbert or Banach spaces has a great importance in nuclear theory and other fields. Eager also applies the integral geometry to modern number theory.

Kinematic measure.

A pair of two orthogonal unit vectors at a point, together with that point is called a frame, and denoted by (Xe_1e_2), hence we have $e_ie_j = \delta_{ij}$ $(i, j = 1, 2)$.

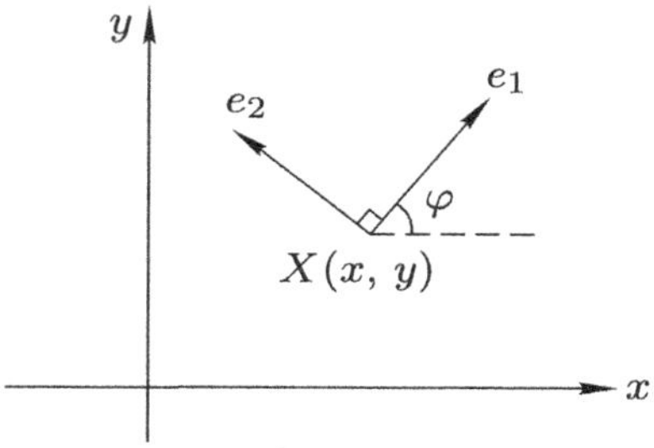

Fig. 22.

To a set of frames we want to introduce a measure, the so-called Kinematic measure. This measure is considered first by Poincaré: And exactly due to this notion that the integral geometry was pushed a great step forward from the classical domains established by Crofton. In the history of integral geometry, many great mathematicians had intimate relationships with it.

In order to determine a frame, we need 2 coordinates (x, y) to fix the origin X, and one more φ to fix e_1. Hence all frames form a three dimensional space. In fact φ can be considered as belonging to S^1, the unit circle, and since X belongs to the two dimensional euclidean plane E, the space of all frames is topologically equivalent to $E \times S^1$.

Next time we shall introduce the kinematic density as

$$\mathscr{K} = dx \wedge dy \wedge d\varphi$$

where (x,y) is the coordinates of X, φ is the angle inclination of e_1 from the x axis. And we shall prove that K has a lot of good properties.

3.3 Lecture III

{The space of all frames} $= E \times S^1$, hence we can choose (x,y,φ), where φ is determined up to 2π, as the coordinates of a frame. A measure is defined to be a triple integral with respect to the coordinates. We define Kinematic measure $= \iiint_D dxdyd\varphi$ where D is a domain in $E \times S^1$, and Kinematic density $X = dx \wedge dy \wedge d\varphi$.

We shall prove that the kinematic measure satisfies three invariance properties.

i) Invariance under motion

A motion of points may be viewed as a coordinate transformation of a fixed point. Thus its equation is represented by:

$$\begin{cases} x = a + x^* \cos\alpha - y^* \sin\alpha \\ y = b + x^* \sin\alpha + y^* \cos\alpha \\ \varphi = \varphi^* + \alpha \end{cases}$$

with constant a, b, α.

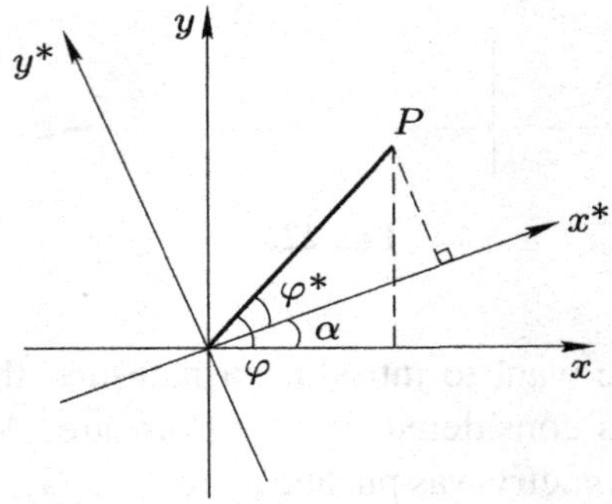

Fig. 23.

Since we always take the absolute value in integrating, it is true that in order to prove the invariance of the kinematic measure, we need only have to prove that the Jacobian of the transformation is either 1 or -1, that is we have to prove the invariance of $\mathscr{K}$ up to a sign.

Now

$$\begin{cases} dx = \cos\alpha dx^* - \sin\alpha dy^*, \\ dy = \sin\alpha dx^* + \cos\alpha dy^*, \\ d\varphi = d\varphi^*, \end{cases}$$

thus

$$dx \wedge dy \wedge d\varphi = (\cos^2\alpha + \sin^2\alpha) dx^* \wedge dy^* \wedge d\varphi^* = dx^* \wedge dy^* \wedge d\varphi^*$$

This proves the result.

ii) Invariance of choice

In order to understand this and to have a clear distinguish from i), we make the following meditations: The position of a rigid figure F (domain, segment, curve, etc.) is determined on the plane by the position of a point $X(x,y)$ of F and the angle φ formed by a direction e_1 fixed in F and a fixed direction O_x of the plane. Hence the figure has coordinates (x,y,φ). This kind of coordinatization of the figure is just equivalent to associating a frame to each figure. And the kinematic measure of a set of figures is just defined to be the kinematic measure of the set of frames associating to them.

Now the invariance of choice means that, if in order to determine the position of F, we take another point X' and another direction e_1' fixed in F, and kinematic measure does not change. Restate more exactly, we have: To each frame $X'e_1e_2$ (associated to a figure F) we take another new frame $X'e_1'e_2'$ (associating to the same figure F) which is rigidly fixed to Xe_1e_2, let D (D' resp.) denotes the domain of the set of frames of Xe_1e_2 ($X'e_1'e_2'$ resp.). Then the measure of frames Xe_1e_2 over D is equal to the measure of frames $X'e_1'e_2'$ over D'.

Denote call (a,b,φ_0) the coordinates of the new frames at X', with respect to the frame Xe_1e_2 at X.

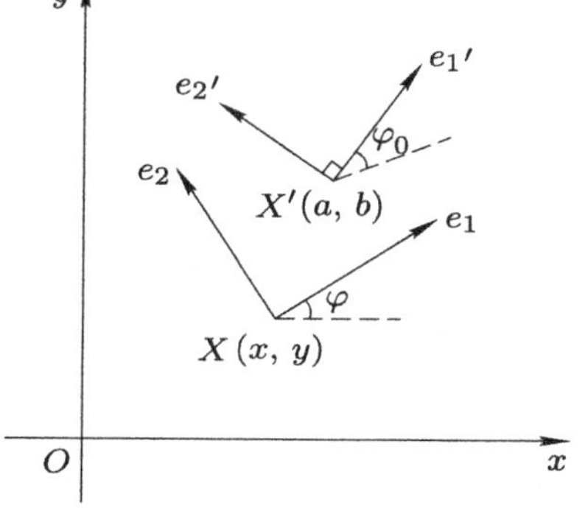

Fig. 24.

We have:

$$\begin{cases} \varphi' = \varphi + \varphi_0, \\ x' = x + a\cos\varphi - b\sin\varphi, \\ y' = y + a\sin\varphi + b\cos\varphi, \end{cases}$$

where (x', y', φ') is the coordinates of $X'e_1'e_2'$ is the system Oxy. Therefore $d\varphi' = d\varphi$, $dx' = dx + (\cdots)d\varphi$, $dy' = dy + (\cdots)d\varphi$,

$$dx' \wedge dy' \wedge d\varphi' = dx \wedge dy \wedge d\varphi$$

which proves the property announced.

iii) Invariance under inversion

For a fixed coordinate system Oxy, we can coordinatize the set of frames Xe_1e_2, and constitute a domain D in $E \times S^1$. As Xe_1e_2 varies in D, it may be regarded as moving in D (referring to the fixed Oxy), this motion differs from the one considered in mechanics (which has one parameter t) only in that it depends on three parameters. Suppose now that an observer fixed with Xe_1e_2 makes some observations, then he does not feel the motion of Xe_1e_2, but rather feels that the frame Oxy is moving. As Xe_1e_2 traces over D, to this observer, the frame Oxy ranges over a domain D_1. The phenomenon is called the inversion of the motion. The coordinates (x_1, y_1, φ_1) of Oxy with reference to Xe_1e_2 is represented by:

$$\begin{cases} \varphi_1 = -\varphi, \\ x_1 = (-x)\cos(-\varphi) - (-y)\sin(-\varphi) = -x\cos\varphi - y\sin\varphi, \\ y_1 = (-x)\sin(-\varphi) + (-y)\cos(-\varphi) = x\sin\varphi - y\cos\varphi. \end{cases}$$

Therefore

$$\begin{aligned} dx_1 &= -\cos\varphi dx - \sin\varphi dy + (\cdots)d\varphi, \\ dy_1 &= \sin\varphi dx - \cos\varphi dy + (\cdots)d\varphi, \\ d\varphi_1 &= -d\varphi, \end{aligned}$$

thus,

$$dx_1 \wedge dy_1 \wedge d\varphi_1 = -(\cos^2\varphi + \sin^2\varphi) dx \wedge dy \wedge d\varphi = -dx \wedge dy \wedge d\varphi.$$

Since in integration we always take the absolute value of the Jacobian, hence the measure is invariant. This establishes the asserting invariance.

We note that ii) is in fact a consequence of i) and iii). Similarly iii) is a consequence of i) and ii), for example, let us show the first: The invariance under motion i) signifies the invariance under a change of fixed coordinate system; this asserts the invariance under a change of mobile coordinate system ii) which is the fixed system in the inverse motion iii).

One who studies the topological groups knows that if a topological group satisfies a certain condition, then there exists a Haar measure, this is perhaps the most fundamental existence theorem in analysis. (See, Pontrjagin, Topological group.)

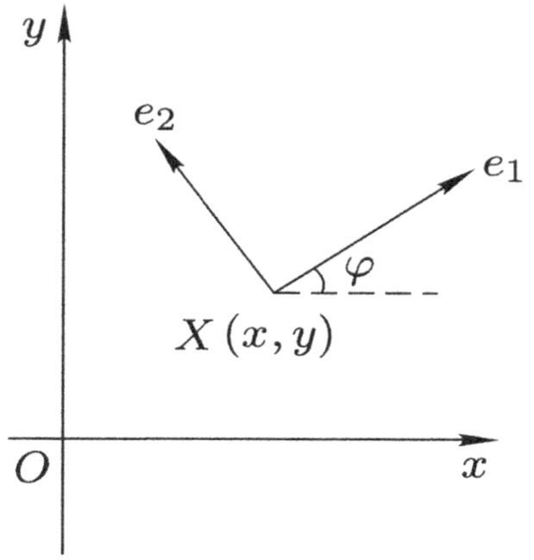

Fig. 25.

Since the space of frames is isomorphic to the group G of proper motions, the kinenatic measure just corresponds to the Haar measure. Moreover we have the following properties-correspondences in the two kinds of measures.

i) $\xleftrightarrow{\text{(correspond)}}$ left invariance, i.e. invariance under the left translation $x \longrightarrow ax$.
ii) $\longleftrightarrow$ right invariance, i.e. under $x \longrightarrow xa$.
iii) $\longleftrightarrow$ invariance under inversion, i.e. under the transformation $x \longrightarrow x^{-1}$.

A Haar measure always has the left invariance property, but in general, it does not have the right invariance property. For some special topological groups, right invariance property holds, it can be proved that, under the left invariance property, the other two invariance properties area equivalent, as we have seen already in the kinematic measure. A Lie group is a very special kind of topological group, so that in it we do not need the conception of a Haar measure.

In order to calculate explicitly the kinematic measure, we have to discuss first the various forms of the kinematic density K in various coordinate systems.

Given a frame Xe_1e_2, it determines a line (p,θ) containing X and has e_1 as its direction, let t denote the distance from X to N, the foot of perpendicular line from O. Then (p,θ,t) determines completely the frame Xe_1e_2 and can be used as the coordinates. We have

$$\begin{cases} x = p\cos\theta + t\sin\theta \\ y = p\sin\theta - t\cos\theta \\ \varphi = \theta - \frac{\pi}{2} \end{cases} \quad \text{thus} \quad \begin{cases} dx = \cos\theta dp + \sin\theta dt + (\cdots)d\theta \\ dy = \sin\theta dp - \cos\theta dt + (\cdots)d\theta \\ d\varphi = d\theta \end{cases}$$

therefore

$$\begin{aligned} dx \wedge dy \wedge d\varphi &= -(\cos^2\theta + \sin^2\theta) dp \wedge dt \wedge d\theta \\ &= dt \wedge dp \wedge d\theta \\ &= dt\mathfrak{y}. \end{aligned}$$

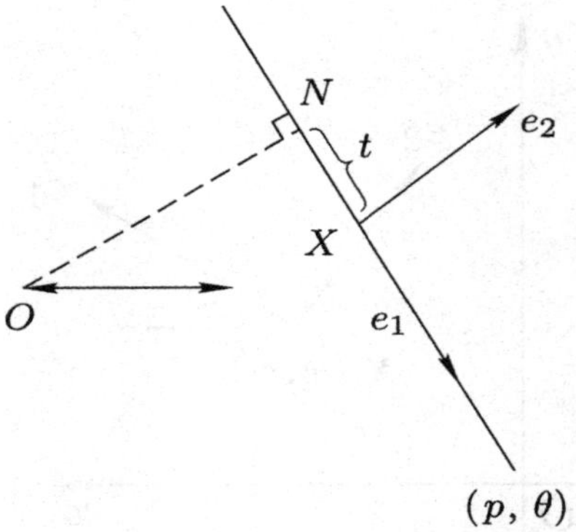

Fig. 26.

Let K be a convex domain with area A, boundary ∂K, and the length of ∂K is L. Let σ be an arbitrary oriented segment of length ℓ. We want to find the kinematic measure over the domain of all frames such that the frame Xe_1e_2 with σ replacing e_1, i.e. with ℓe_1 replacing e_1 has this σ intersecting with K, that is, of all frames such that the σ issuing from their origins Intersect with K.

$$\int_{K\cap\sigma\neq\emptyset}\mathscr{K}\int_{K\cap\sigma\neq\emptyset} dt\mathfrak{y} = \int_{h\cap K\neq\emptyset}\left[\int_A^B dt\right]\mathfrak{y} = \int_{h\cap K\neq\emptyset}(\ell+s)\mathfrak{y},$$

where $s =$ length of the intersection of K with the oriented line h, therefore we have

$$\begin{aligned}\int_{K\cap\sigma\neq\emptyset}\mathscr{K} &= \ell\int_{\substack{h\cap K\neq\emptyset\\ \text{oriented}}}\mathfrak{y}+\int_{\substack{h\cap K\neq\emptyset\\ \text{oriented}}} s\mathfrak{y}\\ &= 2\ell L+2\pi A,\end{aligned}$$

since $\int_{h\cap K\neq\emptyset}\mathfrak{y} = L$ for unoriented lines h, if we consider its orientation, then each

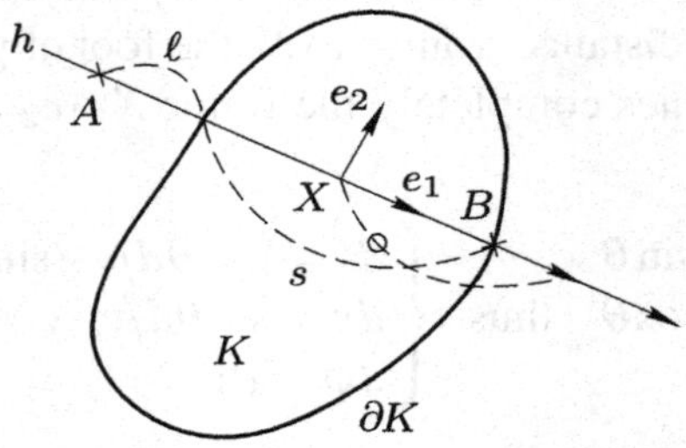

Fig. 27.

unoriented line must be counted twice, therefore $\int_{\substack{h\cap K\neq\emptyset\\ \text{oriented}}}\mathfrak{y} = 2L$, similar argument for the integral $\int s\mathfrak{y}$. Thus we have proved

$$\int_{\substack{K\cap\sigma\neq\emptyset \\ \text{oriented}}} \mathscr{K} = 2\ell L + 2\pi A.$$

from which we obtain $\displaystyle\int_{\substack{K\cap\sigma\neq\emptyset \\ \text{unoriented}}} \mathscr{K} = \ell L + \pi A.$

If K is itself a line segment of length ℓ', then $A = 0$, and L must be regarded as $2\ell'$, therefore we have $\int_{\substack{K\cap\sigma\neq\emptyset \\ \text{oriented}}} \mathscr{K} = 4\ell\ell'$.

If K, σ are polygonal lines of length ℓ', ℓ then $\int_{K\cap\sigma\neq\emptyset} n\mathscr{K} = 4\ell\ell'$, where n is the number of points in $\sigma \cap K$.

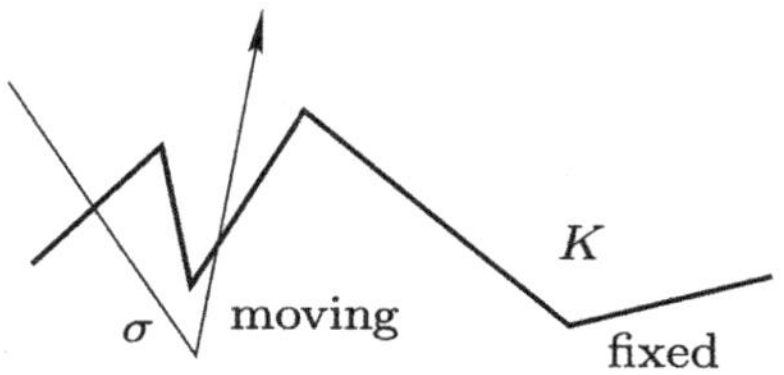

Fig. 28.

According to the invariance under inversion, we may also regard σ as fixed and K moving, the same relation $\int_{K\cap\sigma\neq\emptyset} n\mathscr{K} = 4\ell\ell'$ holds. This formula is called the Formula of Poincaré. We shall give another proof of it.

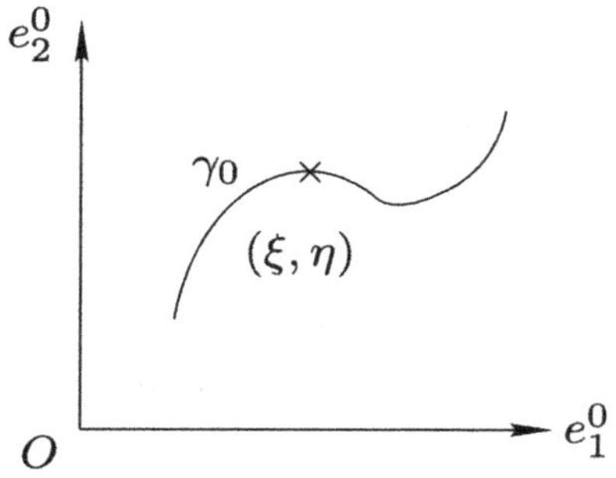

Fig. 29.

A fixed smooth γ_0 has equation $X = \xi(s)$, $Y = \eta(s)$, s arc length referring to a fixed frame $Oe_1^0e_2^0$.

Now let us consider a moving smooth curve γ, and associate to this moving curve a moving frame Xe_1e_2, which is rigidly fixed to the curve. In the coordinate system Xe_1e_2, γ has equations expressed by $X = u(\sigma)$, $Y = v(\sigma)$, σ: arc length.

We want to calculate the kinematic measure over all frames such that the curves γ_0 and γ have intersection points.

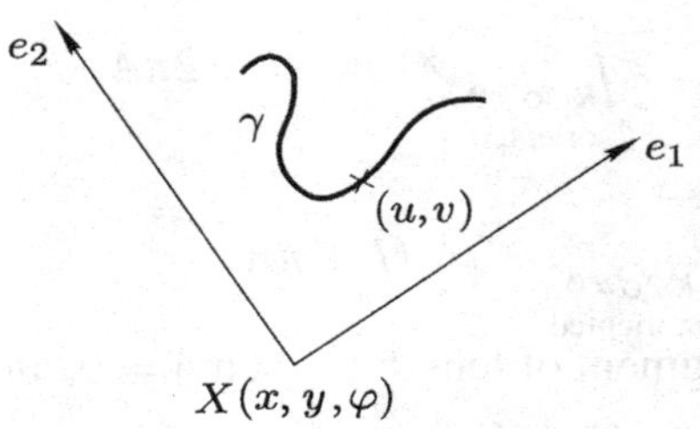

Fig. 30.

We have to express the equations of γ_0, γ in the same, coordinate system. The equations of γ relative to the fixed frame $Oe_1^0e_2^0$ are

$$\begin{cases} X = x + u\cos\varphi - v\sin\varphi, \\ Y = y + u\sin\varphi + v\cos\varphi, \end{cases}$$

where (x, y, φ) is the coordinates of the origin of Xe_1e_2 relative to $Oe_1^0e_2^0$. The common points of γ, γ_0 are given by

$$\begin{cases} \xi(s) = x + u(\sigma)\cos\varphi - v(\sigma)\sin\varphi, \\ \eta(s) = y + u(\sigma)\sin\varphi + v(\sigma)\cos\varphi, \end{cases} \tag{1}$$

from which we can solve for s, σ; in order to determine the pair of arcs γ_0 and γ intersecting at $P = (s, \sigma)$, we have to specify the angle φ of motion of the arc γ. By the invariance property of choice, we can take (s, σ, φ) as the coordinates of a new frame $P\bar{e}_1\bar{e}_2$ rigidly associated to γ which intersects with γ_0 at (s, σ), and is characterized by a motion angle φ, $\frac{d\xi}{ds} = \cos\tau_0$, $\frac{d\eta}{ds} = \sin\tau_0$, τ_0 is the angle of inclination of γ_0, $\frac{du}{d\sigma} = \cos\tau$, $\frac{dv}{d\sigma} = \sin\tau$, τ is the angle of inclination of γ (of course with respect to the frame $P\bar{e}_1\bar{e}_2$ which is rigidly associated to γ. By differentiating (1), we have

$$\begin{aligned} -dx &= -d\xi + \cos\varphi du - \sin\varphi dv + (\cdots)d\varphi \\ &= -\cos\tau_0 ds + (\cos\varphi\cos\tau - \sin\varphi\sin\tau)d\sigma + (\cdots)d\varphi \\ &= -\cos\tau_0 ds + \cos(\varphi+\tau)d\sigma + (\cdots)d\varphi, \\ -dy &= -d\eta + \sin\varphi du + \cos\varphi dv + (\cdots)d\varphi \\ &= -\sin\tau_0 ds + (\sin\varphi\cos\tau + \cos\varphi\sin\tau)d\sigma + (\cdots)d\varphi \\ &= -\sin\tau_0 ds + \sin(\varphi+\tau)\sigma + (\cdots)d\varphi, \end{aligned}$$

thus

$$\begin{aligned} &dx \wedge dy \wedge d\varphi \\ &= -(\cos\tau_0\sin(\varphi+\tau) - \sin\tau_0\cos(\varphi+\tau))ds \wedge d\sigma \wedge d\varphi \\ &= -\sin(\varphi+\tau-\tau_0)ds \wedge d\sigma \wedge d\varphi \\ &= -\sin\omega ds \wedge d\sigma \wedge d\omega. \end{aligned}$$

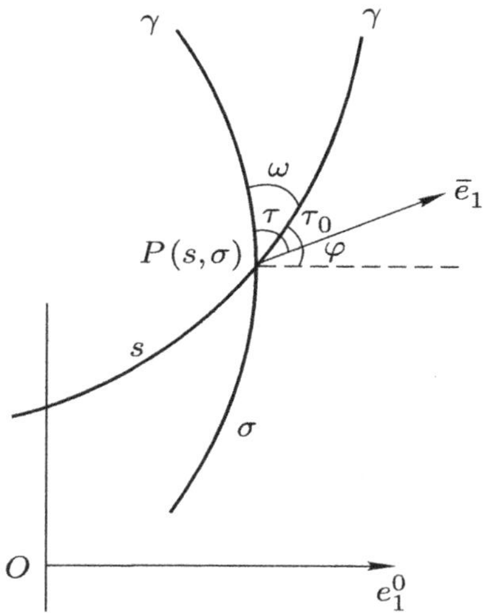

Fig. 31.

Since ω is the angle between the two curves γ_0 and γ, we have

$$\varphi + \tau - \tau_0 = \omega, \quad \text{thus} \quad d\omega = d\varphi.$$

But $\int_0^{2\pi} |\sin \omega| d\omega = 4 \int_0^{\frac{\pi}{2}} \sin \omega d\omega = 4$, if $n =$ the number of common points of $\gamma_0 \wedge \gamma$, we have

$$\begin{aligned} \int n\mathscr{K} &= \int_{\forall (s,\sigma)} \left\{ \int_0^{2\pi} |\sin \omega| d\omega \right\} ds \wedge d\sigma \\ &= 4\ell_0 \ell, \end{aligned}$$

where ℓ_0 is the length of γ_0, and ℓ is the length of γ. Note that $\int n\mathscr{K}$ is integrated over all straight lines, since $n = 0$ if $\gamma \wedge \gamma_0 = \emptyset$.

This formula still holds true in case γ_0, γ are rectifiable curves, but it requires some complicated calculations to approximate γ_0, γ at first.

As a matter of fact, most of the formulas mentioned in these lectures can be extended to more general cases. The problem concerning how far they can be extended is rather complicated, I do not feel that this problem has great importance. H. Federer, W. Maak had some papers discussing this problem. But it has no much interest.

Next time we shall prove the isoperimetric inequality $L^2 - 4\pi A \geq 0$, and Blaschke's kinematic formula, which is the most fundamental formula in integral geometry.

3.4 Lecture IV

A smooth arc γ of length ℓ is fixed, whereas a smooth arc γ' of length ℓ' is moving. By Poincaré's formula we have $\int n(\gamma \wedge \gamma')\mathscr{K}(\gamma') = 4\ell\ell'$ where $n(\gamma \wedge \gamma')$ is the number of intersection points of γ and γ', $\mathscr{K}(\gamma')$ is the kinematic density of the arc γ'.

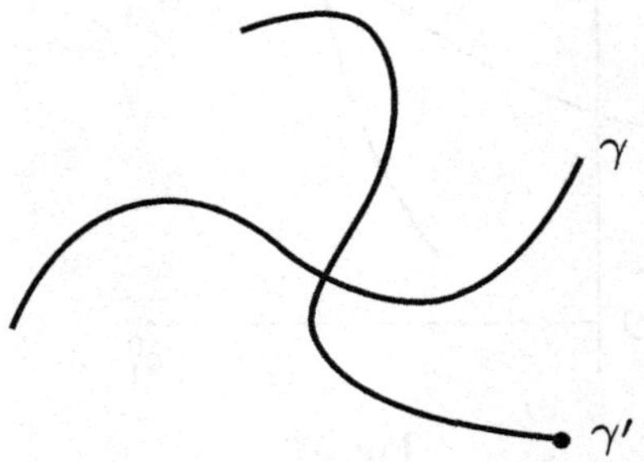

Fig. 32.

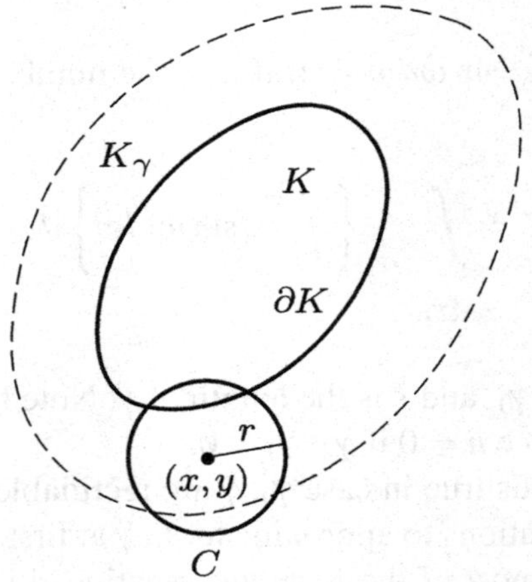

Fig. 33.

Santalo's proof of isoperimetric inequality

Consider a convex domain K of area A, whose boundary ∂K is of length L. If C is a moving circle with (x,y) as center and is of radius γ. Integrating with respect to the kinematic density of G over all circles, we have, applying Poincaré's formula

$$\int n\mathscr{K}(C) = 4L \cdot 2\pi\gamma = 8\pi L\gamma.$$

Fix the center (x,y) for a moment, consider all circles of radius γ with (x,y) as center, (they are distinguished from each other by different φ). On account of the invariance of choice of the frame, we see that they are all equivalent to one another, that is

$$\int_n \mathcal{K}(C) = \int ndxdy\left\{\int_0^{2\pi} d\varphi\right\} = 2\pi\int_{(x,y)=\text{center of } C} ndxdy.$$

Hence we have

$$\int_{(x,y)=\text{center of } C} ndxdy = 4\gamma L.$$

The positions with n odd constitute a set of zero measure, because they are positions of contact, determined by two parameters (set of dimension 2) and the kinematic measure refers to sets of dimension 3.

Therefore we may denote formally

$$\int ndxdy = 2F_2 + 4F_4 + 6F_6 + \cdots,$$

where F_{2k} = the area of domain D_{2k} such that every position in D_{2k} has $2k$ intersection points with K, naturally, we have $F_{2k} \geq 0$. And

$$F_2 + 2F_4 + 3F_6 + \cdots = 2\gamma L. \tag{1}$$

If we set

$$F = F_2 + F_4 + F_6 + \cdots \tag{2}$$

then F represents the area of the domain D such that a circle with radius γ whose center is in the position in D meets ∂K.

Let $K_\gamma = \{\xi\,|\text{dist}(\xi, K) \leq \gamma\}$, from Steiner formula, which will be proved later, the area A_γ of K_γ can be expressed by

$$A_\gamma = A + \gamma L + \gamma^2\pi.$$

Let γ_i be the largest inscribed circle, γ_e the smallest circumscribed circle about K, of course we have $\gamma_i \leq \gamma_e$. If γ is chosen so that $\gamma_i \leq \gamma \leq \gamma_e$, then any circle with radius γ and center in K_γ must meet ∂K, since otherwise it must be contained wholly in K or contains K wholly with no intersection of their boundaries, hence γ either $< \gamma_i$ or $> \gamma_e$, a contradiction. Hence for $\gamma_i \leq \gamma \leq \gamma_e$, $K_\gamma = D$, and $A_\gamma = F$, thus

$$F = A + \gamma L + \gamma^2\pi. \tag{3}$$

From (1), (2),

$$2\gamma L - F = F_4 + 2F_6 + \cdots.$$

From (3)

$$\begin{aligned}2\gamma L - F &= \gamma L - A - \pi\gamma^2\\ &= \left(\frac{L^2}{4\pi} - A\right) - \pi\left(\frac{L}{2\pi} - \gamma\right)^2,\end{aligned}$$

hence

$$\frac{L^2}{4\pi} - A = \pi\left(\frac{L}{2\pi} - \gamma\right)^2 + F_4 + 2F_6 + \cdots \geqslant 0. \tag{4}$$

This is called the equation of Santalo, from which we have the isoperimetric inequality $L^2 - 4\pi A \geq 0$.

Suppose further that K satisfies the condition $L^2 - 4\pi A = 0$, then every term in (4) must vanish, especially $\gamma = \frac{L}{2\pi}$, this means that only one particular value $\gamma = \frac{L}{2\pi}$ can make (4) true, i.e., γ_i must be equal to γ_e. From this we conclude that K must itself be a circle, since it is sandwiched between two circles of the same radius. This proves that the equality sign, holds if and only if K is a circle. From this we have the familiar result. If the length of the boundary is fixed, then a circle has greatest area.

We have a refinement of the isoperimetric inequality, called *Bonnesen's inequality*.

Since (4) holds for all $\gamma_i \leq \gamma \leq \gamma_e$, we have in particular

$$\frac{L^2}{4\pi} - A \geq \pi\left(\frac{L}{2\pi} - \gamma_i\right)^2$$

and

$$\frac{L^2}{4\pi} - A \geq \pi\left(-\frac{L}{2\pi} + \gamma_e\right)^2$$

from the inequality $\frac{1}{2}(x_1^2 + x_2^2) \geq (\frac{x_1+x_2}{2})^2$

$$\frac{L^2}{4\pi} - A \geq \frac{\pi}{2}\left[\left(\frac{L}{2\pi} - \gamma_i\right)^2 + \left(-\frac{L}{2\pi} + \gamma_e\right)^2\right] \geq \pi\left[\left(\frac{\gamma_e - \gamma_i}{2}\right)^2\right]$$

hence $L^2 - 4\pi A \geq \pi^2(\gamma_e - \gamma_i)^2$, which is called the Bonnesen's inequality.

Blaschke's kinematic formula

$D_0 =$ a fixed domain, bounded by a smooth curve ∂D_0, $D =$ a moving domain, bounded by a smooth curve ∂D, the domains need not be simple. $\chi(D_0 \cap D)$: the Euler characteristic of the intersection $D_0 \cap D$. (Note that the Euler characteristic of a domain is defined to be the number equal to the sum of vertices and faces minus the sides, if this domain is a singular complex, e.g. a triangular has Euler characteristic $= 3 + 1 - 3 = 1$.)

Then we have

$$\int_{\text{all positions of } D} \chi(D_0 \cap D)\mathscr{K}(D) = 2\pi(X_0 A + X A_0) + L_0 L$$

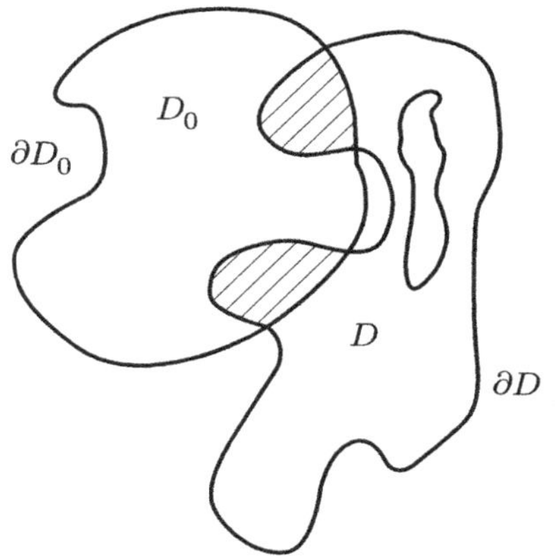

Fig. 34.

where $\mathscr{K}(D)$ is the kinematic density of D, (which can be obtained by considering a frame rigidly attached to D), $A_0(A)$: area of D_0 (D resp.), $L_0(L)$ length of ∂D_0 (∂D resp.), $\chi_0(\chi$ resp.): Euler characteristic of D_0 (D resp.).

We shall prove it later. However let us note first that this formula is symmetric with respect to D and D_0, this necessary must be so, since the kinematic measure is invariant under inversion.

Applications

1) Suppose D_0, D are all convex, the $\chi(D) = \chi_0(D_0) = 1$, $\chi(D \cap D_0) = 1$ if $D \cap D_0 \neq \emptyset$, therefore we have the formula of Santalo

$$\int_{D_0 \cap D \neq \emptyset} \mathscr{K}(D) = 2\pi(A_0 + A) + L_0 L.$$

2) D_0: convex, D: a circle of radius γ, then $L = 2\pi\gamma$, $A = \pi\gamma^2$ therefore

$$\int_{D_0 \cap D \neq \emptyset} \mathscr{K}(D) = 2\pi(A_0 + \pi\gamma^2) + L_0 2\pi\gamma = 2\pi(A_0 + \gamma L + \gamma^2\pi).$$

But the left hand side is equal to

$$\int_{D_0 \cap D \neq \emptyset} dx dy d\varphi = \int_{D_\gamma} \left\{ \int_0^{2\pi} d\varphi \right\} dx dy = 2\pi A_\gamma,$$

where D_γ is the domain parallel to D_0 at distance γ, i.e. $D_\gamma = \{\xi | \mathrm{dist}(\xi, D_0) \leq \gamma\}$ and A_γ is the area of D_γ. Hence

$$A_\gamma = A_0 + \gamma L + \gamma^2 \pi.$$

This is the Steiner's formula which we have acquainted with in proving Poincaré's formula.

3) If D_0 and D are congruent convex domains, from (1), we have

$$\int_{D_0 \cap D \neq \emptyset} \mathcal{K}(D) = 4\pi A + L^2.$$

4) *Another proof of isoperimetrie inequality*

Let $J_k \geq 0$ be the measure of all positions of D such that ∂D_0 and ∂D (where D_0 is congruent to D) have exactly k common points. Then $J_k = 0$ if k is odd, since they are positions of contact. From (3) we have

$$4\pi A + L^2 = J_2 + J_4 + J_6 + \cdots,$$

by Poincaré's formula, since D, D_0 are congruent

$$4L^2 = \int n\mathcal{K}(D) = 2J_2 + 4J_4 + 6J_6 + \cdots$$

$$\therefore \quad 2L^2 = J_2 + 2J_4 + 3J_6 + \cdots$$

thus

$$L^2 - 4\pi A = J_4 + 2J_6 + \cdots \geq 0$$

this proves the inequality.

This equation has another important implication:

If D_0 is not a circle, then $L^2 - 4\pi A = J_4 + 2J_6 + \cdots > 0$ must hold, that is, there exists at least one $J_{2k} > 0$ for $k = 2, 3, \ldots$.

Write it formally, we have the following proposition:

If γ is a closed, convex curve which, is not a circle, then there is a set of positions of positive measure such that it intersects itself in at least four points.

5)

Theorem 3.4.1. *Let D_0, D are simply connected domains bounded by simple closed smooth curves (D_0, D are not necessary convex).*

If, $2\pi(A_0 + A) - L_0 L > 0$, then either $D_0 \subset D$ or $D \subset D_0$.

Proof.: ν: the number of components of $D_0 \cap D$ then $\chi(D_0 \cap D) = \nu$, $\chi = 1$, $\chi_0 = 1$, since D, D_0 are all simply connected. Therefore

$$\int \nu \mathcal{K}(D) = 2\pi(A + A_0) + L_0 L \qquad \text{(Blaschke)}.$$

Moreover we have

$$\int n\mathcal{K}(D) = 4L_0 L \qquad \text{(Poincaré)}.$$

If $D_0 \subset D$ or $D \subset D_0$, consider the number of $\partial D_0 \cap \partial D$, we must have $n = 0$, $\nu = 1$.

Let M_0 be the measure of all positions such that $D \subset D_0$ or $D_0 \subset D$, thus

$$M_0 + \int_{n\neq 0} \nu \mathscr{K}(D) = 2\pi(A + A_0) + L_0 L.$$

If $n \neq 0$, every component of $D \cap D_0$ has at least two intersection points in $\partial D \cap \partial D_0$, that is, $\nu \leq \frac{n}{2}$, or we have

$$\int_{n\neq 0} \nu \mathscr{K}(D) \leq \frac{1}{2} \int_{n\neq 0} n\mathscr{K}(D) = 2L_0 L.$$

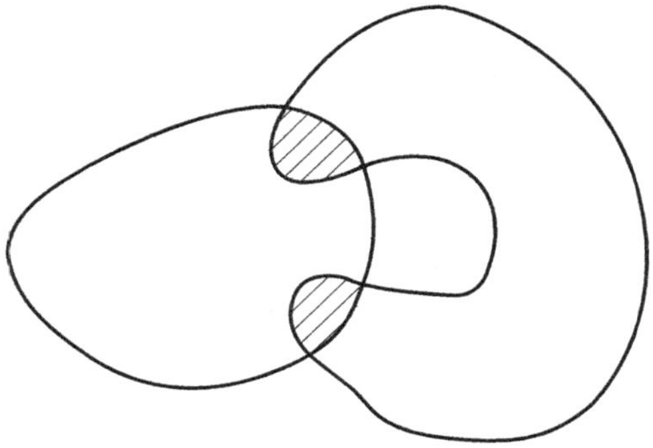

Fig. 35.

As a consequence $M_0 \geq 2\pi(A_0 + A) - L_0 L > 0$ by hypothesis, $M_0 > 0$ means that the measure of the positions for which $n = 0$, $\nu = 1$ (i.e. $D_0 \subset D$ or $D \subset D_0$) is positive, that is to say, this case actually happens, and D can be embedded in D_0 or conversely.

Theorem 3.4.2. $L_0 L - 4\pi A > \sqrt{L_0^2 L^2 - 16\pi^2 A_0 A} \Rightarrow D \subset D_0$.

Proof. $L_0^2 L^2 - 16\pi^2 A_0 A \geq 0$ holds as a consequence of isorperimetric inequalities for D and D_0.

Squaring the hypothesis,

$$\cancel{L_0^2 L^2} - 8\pi A L_0 L + 16\pi^2 A^2 > \cancel{L_0^2 L^2} - 16\pi^2 A_0 A,$$

that is

$$2\pi(A + A_0) - L_0 L > 0.$$

It follows from Theorem 3.4.1, that either $D \subset D_0$ or $D_0 \subset D$.

Suppose that $D_0 \subset D$, then $A_0 \leq A$, and we have

$$0 \leq \sqrt{L_0^2 L^2 - 16\pi^2 A_0 A} < L_0 L - 4\pi A \leq L_0 L - 2\pi(A + A_0) < 0,$$

a contradiction. This result was due to Hadwiger-Santalo. □

6) D_0 is any convex domain, then it contains a circle of radius at least equal to $\frac{A_0}{L_0}$.

Proof. Let D be a circle of radius R, $L = 2\pi R$, $A = \pi R^2$. The condition of Theorem 3.4.2, for $D \subset D_0$ becomes

$$L_0 \cdot 2\pi R - 4\pi \cdot \pi R^2 > \sqrt{L_0^2 \cdot 4\pi^2 R^2 - 16\pi^2 A_0 \pi R^2}$$
$$= 2\pi R \sqrt{L_0^2 - 4\pi A_0},$$

thus

$$L_0 - 2\pi R > \sqrt{L_0^2 - 4\pi A_0},$$
$$2\pi R < L_0 - \sqrt{L_0^2 - 4\pi A_0}.$$

The convex domain D_0 contains a circle of radius less than or equal to

$$R \leq \frac{1}{2\pi}\left\{L_0 - \sqrt{L_0^2 - 4\pi A_0}\right\}.$$

But

$$L_0 - \sqrt{L_0^2 - 4\pi A_0} = \frac{4\pi A_0}{L_0 + \sqrt{L_0^2 - 4\pi A_0}} \geq \frac{4\pi A_0}{2L_0} = \frac{2\pi A_0}{L_0}.$$

Hence D_0 contains a circle of radius at least equal to $\frac{A_0}{L_0}$. With constant area, the longer L_0, the narrower the shape of domain D_0 must be, hence the circle it contains must become smaller. And we see through this that our result above coincides with our common sense, (cf. G. Grünwald, P. Türan, or Santalo: Rev. Union Argentina, vol. 10, 1945).

Proof of Blaschke's formula

First we want to represent the Euler characteristic of a domain by an integral, by theorem of rotating tangent, if ∂D_0, is a smooth curve, we have

$$\frac{1}{2\pi}\int_{\partial D_0} d\tau = \chi(D_0)$$

where τ is the angle of inclination of the tangent.

In fact, the smoothness of ∂D_0 is not practically applicable, since even if ∂D_0, ∂D were smooth the $\partial(D_0 \cap D)$ might not be smooth. Hence we have to replace the above formula by another which holds true in the case ∂D_0 is sectionally smooth. And we have

$$\chi(D_0) = \frac{1}{2\pi}\left(\int_{\partial D_0} d\tau + \sum_{\text{corners}} \text{exterior angles}\right).$$

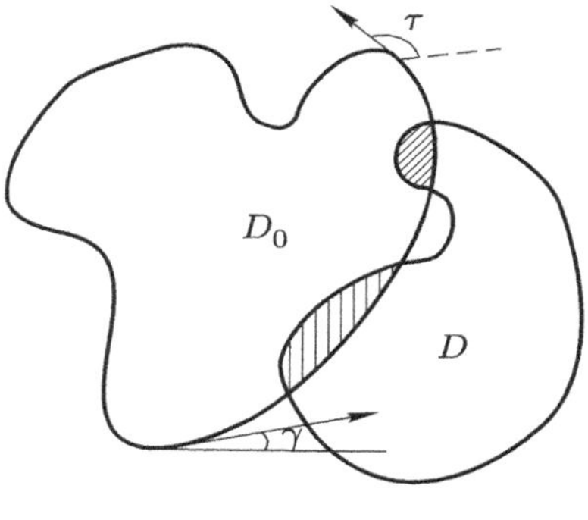

Fig. 36.

This is a special case of the general Gauss-Bonnet theorem, which relates the differential invariants with the topological invariants in a manifold.

Note that since $\partial(D_0 \cap D) = \{\partial D_0 \cap D, \partial D \cap D_0, \text{corners}\}$.

$$2\pi\chi(D_0 \cap D) = \int_{D_0 \cap \partial D} d\tau + \int_{D \cap \partial D_0} d\tau + \sum \text{ext. angle } \psi,$$

D_0 fixed, D moving, we shall consider the integral

$$\int_{\substack{D \cap \partial D_0 \\ \forall D}} d\tau \mathscr{K}(D) \equiv I_1.$$

Let us consider first the measure of the figures congruent to D which contain a fixed point P as an interior point. By the invariance under inversion of the kinematic measure, the result will be the same if we evaluate measure of the points P which are contained in D. In other words $\int_{P \in D} \mathscr{K}(D) =$ measure $(D \ni P)$ = measure $(P \in D) = \int dxdyd\varphi = \int_{P \in D} dxdy \int_0^{2\pi} d\varphi = 2\pi A$, because $P(x,y)$ can vary over the interior of D, and in each position φ can vary from 0 to 2π (where A = area of D).

Return to our I_1, fix for a moment a point P of ∂D_0, then the measure of all positions of D containing this fixed point as an interior point is equal to $2\pi A$. Therefore

$$I_1 = \int_{\substack{\partial D_0 \cap D \\ \forall D}} d\tau \mathscr{K}(D) = \int_{\partial D_0} d\tau \int_{P \in D} \mathscr{K}(D) = 2\pi A \int_{\partial D_0} d\tau = 2\pi A(2\pi\chi_0).$$

By symmetry

$$I_2 = \int_{D_0 \cap \partial D} d\tau \mathscr{K}(D) = 2\pi A_0(2\pi\chi).$$

At least, consider $I_3 = \int_{\text{corners}} \psi \mathscr{K}(D)$. Since $\mathscr{K}(D) = |\sin\psi| d\psi ds d\sigma$, we have

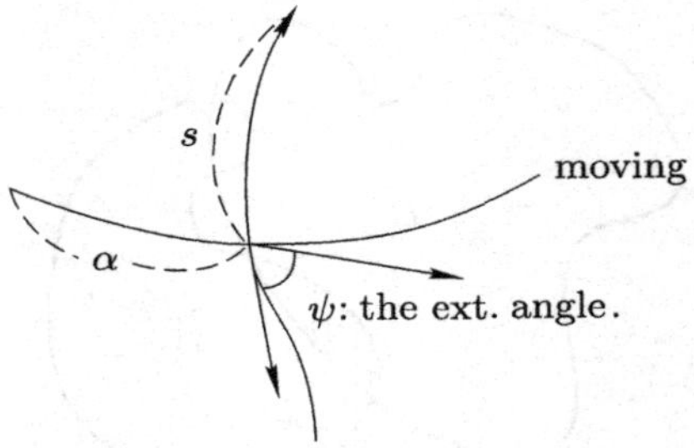

Fig. 37.

$$\begin{aligned} I_3 &= \int |\psi \sin \psi| d\psi ds d\sigma \\ &= \int_0^{L_0} ds \int_0^{L_0} d\sigma \left\{ 2 \int_0^{\pi} (\psi \sin \psi) d\psi \right\}. \end{aligned}$$

But

$$2 \int_0^{\pi} (\psi \sin \psi) d\psi = 2 \left\{ [-\psi \cos \psi]_0^{\pi} + \int_0^{\pi} \cos \psi d\psi \right\} = 2\pi$$

thus

$$I_3 = 2\pi L_0 L.$$

Therefore

$$\begin{aligned} \int \chi(D \cap D_0) \mathscr{K}(D) &= (I_1 + I_2 + I_3) \frac{1}{2\pi} \\ &= (2\pi \chi_0 A + 2\pi \chi A_0) + L_0 L \\ &= 2\pi(\chi_0 A + \chi A_0) + L_0 L. \end{aligned}$$

As we wanted to prove.

This formula is the most fundamental one in integral geometry, Crofton's formula, Poincaré's formula are all special cases or limiting cases of it.

3.5 Lecture V

Integral geometry in euclidean and non-euclidean space of n-dimensions.

In the last four lectures we discussed several interesting results of plane integral geometry, and it seems that it is worthwhile to treat them, because they are meaningful and have some useful applications. We try now to extend our treatments to a high dimensional space, or even a non-euclidean space. This extension complicates the situations, many problems which can be calculated easily in the plane case, become unsolvable in the n-dim. case. This is due to the fact that there are many algebraic difficulties, occurring in this case, which except for a few problems having strong and simple geometric meaning, are very hard to settle. I shall choose some examples, and point out through them the methods used in general integral geometry.

Let E be a euclidean space of n dimension, $E = \{\overrightarrow{x} = (x_1, \ldots, x_n) \in \mathbb{R}^n\}$ with a distance (i.e. a positive definite inner product $(\overrightarrow{x}, \overrightarrow{y})$) defined in it.

In the plane, we defined the measure of a set of lines, in our n-dim. E, we shall define the measure of a set of k-dimensional linear subspaces. Let L be a k-dimensional linear subspace which is completely determined by a system of $(n-k)$ linearly independent linear equations, e.g. a hyperplane is determined by a single equation.

The space of all k-dim. linear subspaces is called the Grassmann manifold $G(n,k)$. The reason of its nomenclature is best seen in the following way. Let $O(n)$ denote the real orthogonal group, of transformations in E. It is a transitive group on the unit $(n-1)$-sphere S^{n-1}. And any element of $O(n)$ carries a k-plane into a k-plane, and, in fact, $O(n)$ is transitive on the subset S of $G(n,k)$, consisting of all L^k passing through the origin. If $L^k \in S$ is a fixed k-plane and L^{n-k} is orthogonal complement, the subgroup of $O(n)$ mapping L^k onto itself splits up into the direct product $O(k) \times O'(n-k)$ of two orthogonal subgroups, the first of which leaves L^{n-k} pointwise fixed and the second leaves L^k pointwise fixed.

It follows that we may identify $S = O(n)/O(k) \times O'(n-k)$, and it is on account of this fact that we may regard S as an analytic manifold, but $G(n,k)$ is obtained from S by translations, hence it is also an analytic manifold, and call it the Grassmann manifold of k-planes in n-space.

Example 3.5.1. Take $n = 3$, $k = 2$, this is the set of planes in 3-space. Any plan is determined by an equation $a_1x_1 + a_2x_2 + a_3x_3 + a_4 = 0$. The equation contains four constants, but the planes determined resp. by $ca_1x_1 + ca_2x_2 + ca_3x_3 + ca_4 = 0$, where c is an arbitrary constant, and the above are actually the same. Hence $\dim G(3,2) = 3$.

Example 3.5.2. $n = 3$, $k = 1$, this is the set of lines in 3-space. Since a line ℓ is wholly determined by the two intersection points A and B of it with two different planes π_1 and π_2, but each of A and B requires two coordinates to fix it. Hence $\dim G(3,1) = 4$. Or, a straight line is determined by the two points on it, each of which requires 3 coordinates. But each point can slide along the line, hence $\dim G(3,1) = (3+3) - 2 = 4$.

In general, for the $G(n,k)$, a k-plane L is determined by $(k+1)$ linearly independent points, hence it requires $n(k+1)$ coordinates. But every point on L has k degree of freedom, hence we have

$$\dim G(n,k) = n(k+1) - k(k+1) = (n-k)(k+1).$$

We have to introduce a measure in the domain $G(n,k)$, which is invariant under the group of motions in the n-space. The equations of L are:

$$\sum_{1 \le k \le n} a_{ik}x_k + b_i = 0 \qquad i = 1, \ldots, n-k.$$

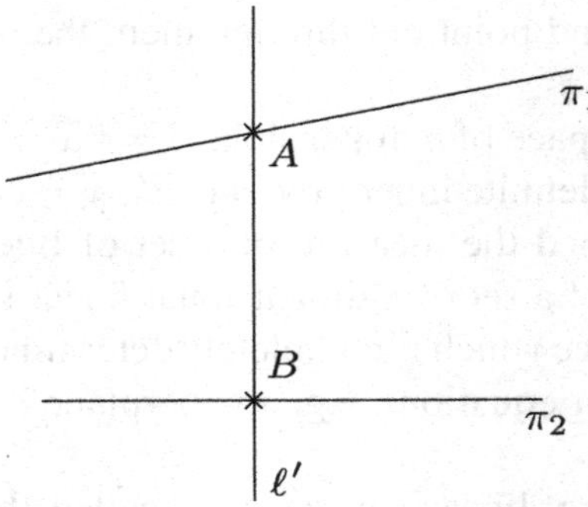

Fig. 38.

How to make a multiple integral of degree dim $G(n,k)$, having the property that it is invariant under motion, from these $(n-k)$ linearly independent equations? It is very difficult to make such a consideration. Hence we have another more effective method by using frames. This is a special case in the treatments of homogeneous space in the Lie group theory.

$(\vec{x}, \vec{e}_1, \ldots, \vec{e}_n)$ is called a frame with x as origin, if $\vec{e}_i \vec{e}_j = \delta_{ij}$, that is, if they are orthonormal vectors. In coordinate forms, we have,

$$\vec{x} = (x_1, x_2, \ldots, x_n), \vec{e}_i = (e_{i1}, e_{i2}, \cdots, e_{in})$$

The differential $d\vec{x} = (dx_1, \ldots, dx_n)$, being a vector, is expressible as the linear combinations of the n independent vectors $\vec{e}_i$, with linear differential forms as coefficients.

$$d\vec{x} = (dx_1, \ldots, dx_n) = \sum_i \omega_i \vec{e}_i \qquad \omega_i : \ell.\ \text{diff. forms.}$$

Multiplying by $\vec{e}_j$ in both sides, we have

$$(\vec{e}_j, d\vec{x}) = \sum_i \omega_i (\vec{e}_i, \vec{e}_j) = \sum_i \omega_i \delta_{ij} = \omega_j,$$

thus

$$\omega_j = \sum_{i=1}^{n} e_{ji} dx_i.$$

Similarly, $d\vec{e}_i = \sum_j \omega_{ij} \vec{e}_j$ with ω_{ij} linear differential forms.

$$(\vec{e}_k, d\vec{e}_i) = \sum_j \omega_{ij} \delta_{jk} = \omega_{ik}$$

thus

$$\omega_{ik} = (\vec{e}_k, d\vec{e}_i) = \sum_\ell e_{k\ell} de_{i\ell}.$$

Now

$$\omega_{ki} = \sum_{\ell} e_{i\ell} de_{k\ell},$$

$$\begin{aligned}\omega_{ik} + \omega_{ki} &= (\vec{e}_k, d\vec{e}_i) + (\vec{e}_i, d\vec{e}_k) \\ &= (\vec{e}_k, d\vec{e}_i) + (d\vec{e}_k, \vec{e}_i) = d(\vec{e}_k, \vec{e}_i) = d(\delta_{ki}) = 0,\end{aligned}$$

hence $\omega_{ik} = -\omega_{ki}$, especially $\omega_{ii} = 0$. Only $n\omega_j$ and $\frac{n(n-1)}{2}$ linearly independent ω_{ik} remain. Their total number is $n + \frac{n(n-1)}{2} = \frac{n(n+1)}{2}$.

The group of motions in E can be generated by translations and rotations, hence they are represented by the system of linear equations:

$$T : x_i' = \sum_k a_{ik} x_k + b_i, \qquad 1 \leq i, k \leq n,$$

with $A = (a_{ik})$ an $n \times n$ orthogonal matrix.

An orthogonal matrix has $n^2 - nC_2 - n = \frac{n(n-1)}{2}$ parameters, therefore the group of motions has $\frac{n(n-1)}{2} + n = \frac{n(n+1)}{2}$ parameters. This number is equal to the independent linear differential forms ω_j and ω_{ik}.

These ω_j, ω_{ik} play a fundamental role in differential geometry. Their importance comes from the fact that they are invariant under motions. This invariance property can be proved in the following.

Consider the motion $T : x_i' = \sum_k a_{ik} x_k + b_i, 1 \leq i, k \leq n$. $A : (a_{ik})$ orthogonal. T maps a point x to a point x', and the unit vectors e_j to e_j'. e_j' is related to e_j by $e_{ji}' = \sum_k a_{ik} e_{jk}$, since, only the rotation A has effect on the vectors. Let us make a little calculation.

$$\begin{aligned}\omega_{ik}' = (\vec{e}_k', d\vec{e}_i') &= \sum_{\ell} e_{k\ell}' de_{i\ell}' \\ &= \sum_{\ell,h,j} (a_{\ell h}, e_{kh})(a_{\ell j}, de_{ij}) \\ &= \sum_{h,j} \delta_{hj}\, e_{kh} de_{ij} \\ &= \sum_j e_{kj} \cdot de_{ij} = \omega_{ik}.\end{aligned}$$

Similarly we can prove $\omega_j' = \omega_j$ under the notion T.

After having these invariant differential forms, we want to prove that we can introduce into $G(n,k)$ a density as the exterior product of some of them.

To each L^k of $G(n,k)$ associate a frame $\vec{x}\,\vec{e}_1 \cdots \vec{e}_n$ such that $\vec{x} \in L$, $\vec{e}_1 \cdots \vec{e}_k \in L^k$, hence $\vec{e}_{k+1} \cdots \vec{e}_n \perp L^k$, and such that the frames associated to various L^k are locally differentiable, therefore we can consider the invariant linear differential forms ω_j, ω_{ik}. Take

$$\mathfrak{y}(n,k) \overset{\text{def}}{=\!=} \bigwedge_{\substack{k+1\leq\gamma,s\leq n\\ 1\leq\alpha\leq k}} \omega_\gamma\omega_{\alpha s} = \omega_{k+1}\wedge\cdots\wedge\omega_n\wedge\omega_{1(k+1)}\wedge\cdots \wedge\omega_{1n}\wedge\omega_{2(k+1)}\wedge\cdots\wedge\omega_{kn}.$$

Its degree is $(n-k)+k(n-k)=(k+1)(n-k)=\dim G(n,k)$.

We shall prove that it is independent of the choice of the frame associated to each L^k. The invariance of $\wedge\omega_\gamma\omega_{\alpha s}$ is not obvious, because we associate another frame to L^k, then ω_j, ω_{ik} will change. One must distinguish between the choice of frames and the motion, they are two different things.

The proof runs as follows:

(note that we have $\begin{matrix}\alpha,\beta,\gamma,\ldots:1,2,\ldots,k\\ \gamma,s,t\ldots:k+1,\ldots,n\end{matrix}$ as a convention.)

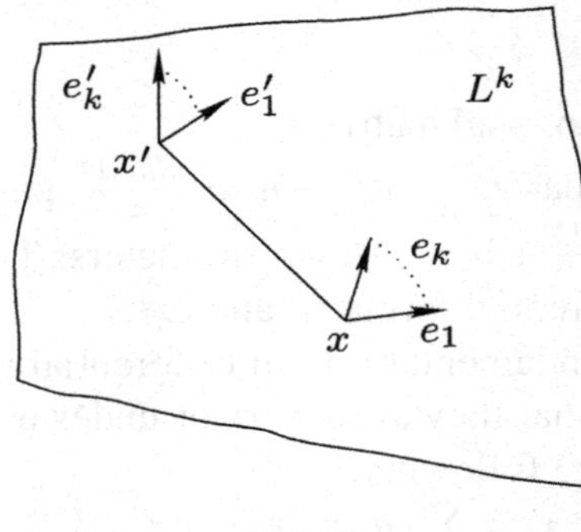

Fig. 39.

Let $\overrightarrow{x}'\overrightarrow{e}'_1\overrightarrow{e}'_2\cdots\overrightarrow{e}'_n$ be another family of frames satisfying the same conditions. Since $\overrightarrow{x}$, $\overrightarrow{x}'\in L^k$, we have $\overrightarrow{x}'-\overrightarrow{x}\parallel L^k$, that is $\overrightarrow{x}'=\overrightarrow{x}+\sum_\alpha u_\alpha\overrightarrow{e}_\alpha$, $\overrightarrow{e}'_\alpha=\sum_\beta u_{\alpha\beta}\overrightarrow{e}_\beta(u_{\alpha\beta})$: a $(k\times k)$ orthogonal matrix.

$\overrightarrow{e}'_r=\sum_s u_{\gamma s}\overrightarrow{e}_s(u_{\gamma s})$: an $(n-k)\times(n-k)$ orthogonal matrix. Therefore,

$$\begin{aligned}\omega'_{\alpha s} &= (de'_\alpha,e'_s)=\left(\left(\sum_\beta du_{\alpha\beta}\overrightarrow{e}_\beta+\sum_\beta u_{\alpha\beta}d\overrightarrow{e}_\beta\right),\overrightarrow{e}'_s\right)\\ &=\left(\sum_\beta u_{\alpha\beta}d\overrightarrow{e}_\beta,\sum_\gamma u_{s\gamma}\overrightarrow{e}_\gamma\right)=\sum_{\beta,\gamma}u_{\alpha\beta}u_{s\gamma}(d\overrightarrow{e}_\beta,\overrightarrow{e}_\gamma)\\ &=\sum_{\beta,\gamma}u_{\alpha\beta}u_{s\gamma}\omega_{\beta\gamma}.\end{aligned}$$

$$\bigwedge_{\alpha,s}\omega'_{\alpha s}=(\det u_{\alpha\beta})^{n-k}(\det u_{\gamma s})^k\bigwedge_{\beta,\gamma}\omega_{\beta\gamma}=\bigwedge_{\beta,\gamma}\omega_{\beta\gamma}. \quad (1)$$

Again,

$$\begin{aligned}\omega'_\gamma &= (d\vec{x}', \vec{e}'_\gamma)\\ &= \left(d\vec{x} + \sum_\alpha \cancel{du_\alpha} \vec{e}_\alpha + \sum_\alpha u_\alpha d\vec{e}_\alpha, \vec{e}'_\gamma\right)\\ &= \left(d\vec{x} + \sum_\alpha u_\alpha d\vec{e}_\alpha, \sum_s u_{\gamma s}\vec{e}_s\right)\\ &= \sum_s u_{\gamma s}\omega_s + \sum(\cdots)\omega_{\alpha s}.\end{aligned}$$

$$\bigwedge_{\alpha,\gamma,s} \omega'_\gamma \omega'_{\alpha s} = \pm\left(\det u_{\gamma s} \bigwedge_{\beta,\gamma,s} \omega_s \omega_{\beta\gamma}\right) = \pm \bigwedge_{\alpha,\gamma,s} \omega_\gamma \omega_{\alpha s}.$$

Hence we have proved that the $\wedge_{\alpha,\gamma,s}\omega_\gamma\omega_{\alpha s}$ differs at most by a sign from that of another frames system. This does not matter, since in considering the measure we always take the absolute value.

This fact justifies our notation above, we define the density of it as the $\mathfrak{y}(n,k) = \wedge\omega_\gamma\omega_{\alpha s}$.

This is a typical example showing how we can make a different approach to avoid the disagreeable algebraic difficulties.

The following example will show that after specializing this general form $\mathfrak{y}(n,k)$, we just obtain the concept of the density of a set of lines previously defined.

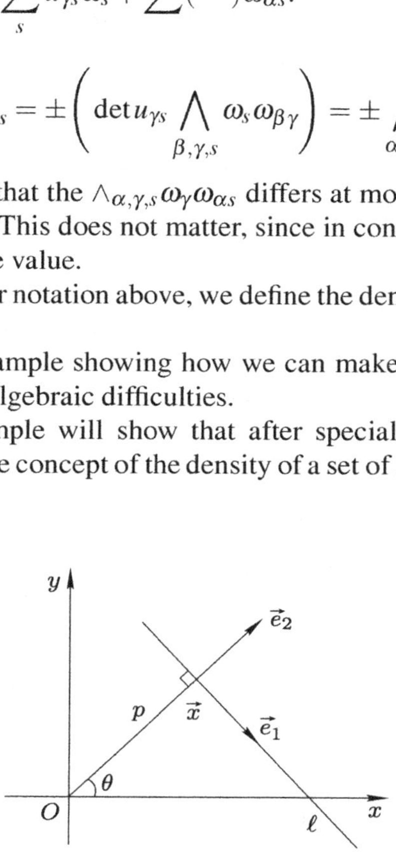

Fig. 40.

Example 3.5.3. $n = 2, k = 1$, $\mathfrak{y}(2,1) = \omega_2 \wedge \omega_{12}$.

Taking $\vec{x}$ to be the foot of the perpendicular line from the origin, the equation of the line becomes

$$x\cos\theta + y\sin\theta = p.$$

Thus

$$\cos\theta dx + \sin\theta dy + (\cdots)d\theta = dp.$$

Let $\vec{e}_2$ run along the direction $O\vec{x}$, therefore $\vec{e}_2 = (\cos\theta, \sin\theta)$, $\vec{e}_1 = (\sin\theta, -\cos\theta)$

$$\omega_{12} = (d\vec{e}_1, \vec{e}_2) = (\cos^2\theta + \sin^2\theta)d\theta = d\theta.$$
$$\omega_2 = (\vec{e}_2, d\vec{x}) = \cos\theta dx + \sin\theta dy = dp - (\cdots)d\theta$$

And so $\mathfrak{y}(2,1) = \omega_2 \wedge \omega_{12} = dp \wedge d\theta$. This result coincides with the previous definition. Some geometric meanings of $\mathfrak{y}(n,k)$ can be clearly pointed out. The $\wedge_{\alpha,s}\omega_{\alpha s}$ is composed of $\omega_{\alpha s}$ which are the components of $d\vec{e}_\alpha$ in the frame $\vec{x}\vec{e}_1\cdots\vec{e}_n$. This means that we make a rotation about $\vec{x}$, i.e. the $\overline{L}^k$ containing the frame $\vec{x}(\vec{e}_1 + d\vec{e}_1)\cdots(\vec{e}_n + d\vec{e}_n)$ is a k subspace containing $\vec{x}$. Hence $\wedge_{\alpha,s}\omega_{\alpha s}$ may be considered as the density of all L^k about a fixed point.

On the other hand, the components ω_γ in $\wedge\omega_r$ are the coefficients of $d\vec{x}$ in the direction $\vec{e}_\gamma$. Hence $\wedge_\gamma\omega_\gamma$ represents the density of all parallel L_k, it is a density of displacement.

The formula $\mathfrak{y}(n,k) = \wedge_\gamma\omega_\gamma \wedge_{\alpha,s}\omega_{\alpha s}$ holds true not only in Euclidean space, but also in non-euclidean space. Moreover the consideration of it in the non-euclidean case is sometimes helpful to that of euclidean case. Because in obtaining the measure, some integrations are very difficult if we don't have the assurance that the space is compact. If we consider the spherical space at the same time, since it is compact, we can make the problems easier. My experience is that it is most convenient to discuss the euclidean and non-euclidean spaces in the same treatment.

Crofton's formula in E

Let M^q be a q-dimensional compact manifold in E.

Consider $G(n,k)$ with k satisfying $k+q-n \geq 0$. Under this assumption, M^q and L^K have in general a $(k+q-n)$-dim, intersection. If $v_q(M^q)$ represents the q-dim volume of M^q, $v_{k+q-n}(L^q \cap M^q)$ represents the $(k+q-n)$-dim volume of the intersection $L^k \cap M^q$, O_m represents the surface area of a unit sphere in m-space, i.e, its $(m-1)$-dim. "area", and assume that $O_1 = 2$. (Note that O_m are absolute constants which can be calculated by gamma function.)

Then the Crofton's formula asserts that:

$$\int v_{k+q-n}(L^k \cap M^q)\mathfrak{y}(n,k) = \frac{O_{n+1}\cdots O_{n-k+1}Q_{k+q-n+1}}{O_1O_2\cdots O_{k+1}O_{q+1}} v_q(M^q).$$

If $k+q-n=0$, then $v_{k+q-n}(L^k \cap M^q) =$ the number v of common points of L^k and M^q. Let C represent the coefficient on the right hand side, then it becomes

$$\int v\vec{(}n,k) = Cv_q(M^q).$$

We find that it has a great similarity to the Crofton's formula in the plane case.

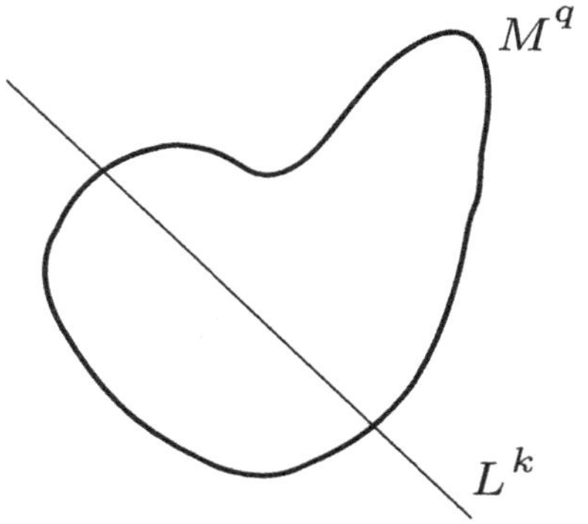

Fig. 41.

Remark 3.5.4. *The formula still holds true if M^q is a more general species of manifold in E, or if E is replaced by a non-euclidean space. First let us illustrate some special cases.*

I: $n = 2$.

$\alpha)$ $k = 1$, $q = 1$, thus $G = \frac{O_3 \not{O}_2 \not{O}_1}{\not{O}_1 \not{O}_2 O_2} = \frac{4\pi}{2\pi} = 2$, $\beta)$, $k = 1$, $q = 2$, $C = \frac{\not{O}_3 \not{O}_2 O_2}{O_1 \not{O}_2 \not{O}_3} = \frac{2\pi}{2} = \pi$, $\int \mathfrak{y}(2,1) = 2L$, thus $\int \mathfrak{y}(2,1) = \pi A$.

II: $n = 3$.

$$\alpha) \qquad k = 1, q = 2, \text{thus } C = \frac{O_4 \not{O}_3 \not{O}_1}{\not{O}_1 O_2 \not{O}_3} = \frac{2\pi^2}{2\pi} = \pi.$$

$$\beta) \qquad k = 1, q = 3, \text{thus } C = \frac{\not{O}_4 O_3 \not{O}_2}{O_1 \not{O}_2 \not{O}_4} = 2\pi.$$

$$\gamma) \qquad k = 2, q = 1, \text{thus } C = \frac{O_4 \not{O}_3 \not{O}_2 \not{O}_1}{\not{O}_1 \not{O}_2 \not{O}_3 O_2} = \frac{2\pi^2}{2\pi} = \pi.$$

$$\delta) \qquad k = 2, q = 2, \text{thus } C = \frac{O_4 \not{O}_3 \not{O}_2 O_2}{O_1 \not{O}_2 \not{O}_3 O_3} = \frac{\not{2}\pi^2 \cdot \not{2}\not{\pi}}{2 \cdot \not{4}\not{\pi}} = \frac{\pi^2}{2}.$$

$$\varepsilon) \qquad k = 2, q = 3, \text{thus } C = \frac{\not{O}_4 \not{O}_3 \not{O}_2 O_3}{O_1 \not{O}_2 \not{O}_3 \not{O}_4} = \frac{4\pi}{2} = 2\pi.$$

Note that $O_4 = 2\pi^2$. In general a space in E^4 of radius γ has "surface area" $2\pi^2\gamma^3$.

special case	value of the integral	special case	value of the integral
I: $n = 2$: $\alpha)k = 1, q = 1$	$2L$	II: $n = 3 : \alpha)k = 1, q = 3$	πA
$\beta)k = 1, q = 2$	πA	$\beta)k = 1, q = 3$	$2\pi v$
		$\gamma)k = 2, q = 1$	πL
		$\delta)k = 2, q = 2$	$\frac{\pi^2}{2}A$
		$\varepsilon)k = 2, q = 3$	$2\pi v$

Non-euclidean spaces:

A. *Elliptic or spherical space*

Consider in euclidean space of $n+1$ dimension, the sphere

$$x_1^2+x_2^2+\cdots+x_{n+1}^2=\frac{1}{K}>0.$$

The geometry on this space is elliptic geometry (spherical geometry), with curvature K. Identification of points $\overrightarrow{x}$ and $-\overrightarrow{x}$ gives elliptic space. Locally speaking, a spherical space and an elliptic space are the same, the latter may be considered as a covering space of the former.

B. *Hyperbolic space*

Consider Lorentzian space L of $(n+1)$ dimensions, which is the vector space $(x_1,\ldots,x_{n+1})$ with scalar product

$$(\overrightarrow{x},\overrightarrow{y})_L=x_1y_1+\cdots+x_ny_n-x_{n+1}y_{n+1}.$$

The metric in L is not positive definite, it is of signature $(+\cdots+-)$, hence L is not a Riemannian space. However, if we consider in L the hyperquadratic

$$x_1^2+\cdots+x_n^2-x_{n+1}^2=-\frac{1}{K}\quad(\text{where}\quad K>0).$$

Obviously, $x_{n+1}\neq 0$, hence it contains at least two connected components. In fact, this is a hyperboloid of two sheets, as can be seen by considering

$$-x_1^2-x_2^2+x_3^2=\frac{1}{K}.$$

Take the connected component $x_{n+1}>0$. The geometry on this component is hyperbolic geometry with curvature $-K,(K>0)$.

The metric of the Lorentzian space L (not a Riemannian metric) induces on this component a positive definite metric, hence the geometry on this component is also a Riemannian geometry.

Since two surfaces are regarded as equivalent if they are isometric, we can prove that the only simply connected and complete surfaces with curvature $+K$ or $-K$ are the ones illustrated above.

The density formula $\mathfrak{y}(n,k)$ holds true for all these non-euclidean spaces.

*1) $\omega'_{\alpha s}=\sum_\beta u_{\alpha\beta}(\sum_\gamma u_{s\gamma}\omega_{\beta\gamma})\equiv\sum_\beta u_{\alpha\beta}h_{s\beta}$ for any s,

$$\bigwedge_\alpha \omega'_{\alpha s}=(\det u_{\alpha\beta})\bigwedge_\beta h_{s\beta},$$

therefore

$$\bigwedge_{\alpha,s}=(\det u_{\alpha\beta})^{n-k}\bigwedge_{\beta,s}h_{s\beta}.$$

Fig. 42.

But $h_{s\beta} = \sum_\gamma u_{s\gamma}\omega_{\beta\gamma}$, for any β, we have

$$\bigwedge_s h_{s\beta} = (\det u_{s\gamma}) \bigwedge_\gamma \omega_{\beta\gamma},$$

therefore

$$\bigwedge_{\beta,s} h_{s\beta} = (\det u_{s\gamma})^k \bigwedge_{\beta,\gamma} \omega_{\beta\gamma}.$$

Combine these results, we have

$$\bigwedge_{\alpha,s} \omega'_{\alpha s} = (\det u_{\alpha\beta})^{n-k}(\det u_{\gamma s})^k \bigwedge_{\beta,\gamma} \omega_{\beta\gamma}.$$

*2) In our 3-space, let us consider a sequence of overlapping spheres with decreasing radius, and draw it to the fourth dimension, we obtain the S^3 (of radius γ). Hence it's surface "area" can be obtained by the following consideration.

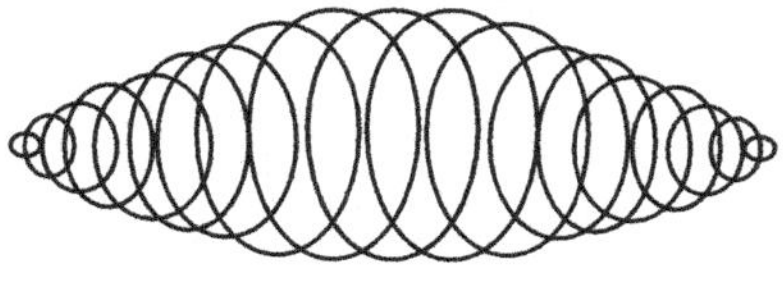

Fig. 43.

In the angle is θ, then the radius of the S^2 sphere (on the surface of our S^3 sphere) is $\gamma \sin\theta$, but the surface area of this S^2 is $4\pi(\gamma\sin\theta)^2$.

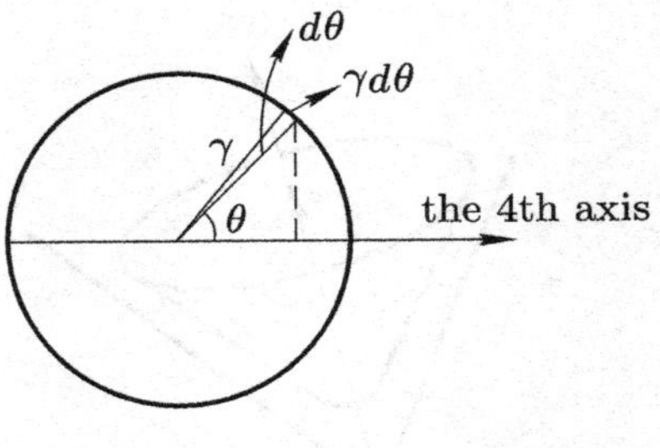

Fig. 44.

Hence the surface area of S^3 with radius γ is

$$SV(s^3) = \int_0^{\pi} 4\pi(\gamma \sin\theta)^2 \cdot \gamma d\theta$$
$$= 4\pi\gamma^3 \int_0^{\pi} \sin^2\theta d\theta.$$

By the formula: $\int \sin^n x dx = -\frac{1}{n}\sin^{n-1} x \cos x + \frac{n-1}{n}\int \sin^{n-2} x dx$

$$SV(S^3) = 4\pi\gamma^3 \cdot \frac{1}{2}\int_0^{\pi} d\theta = 2\pi^2\gamma^3.$$

This is the formula required.

Similarly

$$SV(S^4) = \int_0^{\pi} 2\pi^2(\gamma\sin\theta)^3 \cdot \gamma d\theta = \frac{8}{3}\pi^2\gamma^4, \text{etc.}$$

In general, if n is odd,

$$SV(S^n) = \frac{\pi\gamma^2}{(\frac{n-1}{2})} SV(S^{n-2}),$$
$$SV(S^{n+1}) = SV(S^n) \cdot \gamma \cdot \frac{(n-1)(n-3)\cdots 2}{n(n-2)\cdots 3 \cdot 1} \cdot 2$$
$$= 2\gamma SV(S^n)\frac{(n-1)^2(n-3)^2 \cdots 2^2}{n!}.$$

By the similar method, we can get the volume of S^n.

For example,

$$V(S^2) = \int_0^{\pi} \pi(\gamma \sin\theta)^2 \cdot \gamma \sin\theta d\theta$$
$$= \pi\gamma^3 \int_0^{\pi} \sin^3\theta d\theta = \pi\gamma^3 \cdot \frac{2}{3} \cdot 2 = \frac{4}{3}\pi\gamma^3,$$
$$V(S^3) = \int_0^{\pi} \frac{4}{3}\pi(\gamma \sin\theta)^3 \gamma \sin\theta d\theta$$
$$= \frac{4}{3}\pi\gamma^4 \int_0^{\pi} \sin^4\theta d\theta = \frac{4}{3}\pi\gamma^4 \frac{3}{4}\frac{\pi}{2} = \frac{1}{2}\pi^2\gamma^4., etc.$$

3.6 Lecture VI

We want to prove the Crofton's formula:

$$\int v_{q+k-n}(L^k \cap M^q)\mathfrak{y}(n,k) = Cv_q(M^q),$$

$L^k \cap M^q$ has in general dimension $k+q-n$, it can be proved that the set L^k for which dim $(L^k \cap M^q) > k+q-n$ has measure zero, hence this set need not be accounted in the Crofton's formula.

Consider frames $\overrightarrow{x}\overrightarrow{e}_1\overrightarrow{e}_2\cdots\overrightarrow{e}_n$ such that $x \in L^k \cap M^q$, $\overrightarrow{e}_1,\ldots,\overrightarrow{e}_{q+k-n}$ are tangent vectors to $L^k \cap M^q$ at $\overrightarrow{x}$, and $\overrightarrow{e}_{k+q-n+1},\ldots,\overrightarrow{e}_q$ are tangent vectors to M^q at $\overrightarrow{x}$.

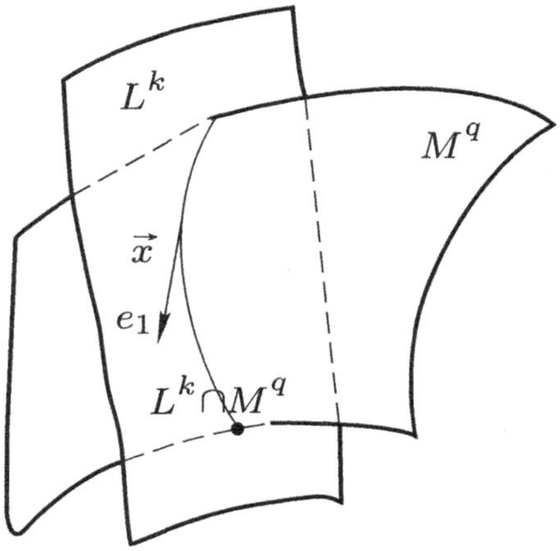

Fig. 45.

Consider another system of frames $\overrightarrow{x}\overrightarrow{f}_1\cdots\overrightarrow{f}_n$ such that $x \in L^k \cap M^q$ and that $\overrightarrow{f}_i = \overrightarrow{e}_i, 1 \le i \le k+q-n$, $\overrightarrow{f}_{q+k-n+1},\ldots,\overrightarrow{f}_k$ are parallel to L^k.

In the cartesian product $G(n,k) \times M^q = \{(L^k,x) | \overrightarrow{x} \in M^q\}$, take a subset

$$S = \{(L^k, \overrightarrow{x}) | \overrightarrow{x} \in L^k\}.$$

When integrate over S, we have two methods corresponding respectively to the two projections from $G(n,k) \times M^q$ to $G(n,k)$ and M^q. Their values, representing the same integral, must be equal, in this way we obtain the required equality formula. All $\overrightarrow{x}$ belong to M^1, therefore $d\overrightarrow{x} = \sum_{i=1}^{q} \omega_i \overrightarrow{e}_i$. If $k+1 \leq h < n$,

$$(d\overrightarrow{x}, \overrightarrow{f}_h) = \sum_{i=1}^{q} \omega_i(\overrightarrow{e}_i, \overrightarrow{f}_h) = \sum_{i=1}^{q+k-n} \omega_i(\overrightarrow{e}_i, \overrightarrow{f}_h) + \sum_{q+k-n+1 \leq i \leq q} \omega_i(\overrightarrow{e}_i, \overrightarrow{f}_h), \tag{1}$$

since in the first term $\overrightarrow{e}_i = \overrightarrow{f}_i$, and $q+k-n < k+1$.

Define $\Phi = \omega_1 \wedge \cdots \wedge \omega_{q+k-n}$, this is the volume element of $L^k \cap M^q$. From (1), we find that $((\overrightarrow{e}_i, \overrightarrow{f}_h))$ is an $(n-k) \times (n-k)$ matrix, where $k+1 \leq h \leq n$ and $k+q-n+1 \leq i \leq q$, hence we have the equality

$$\Phi \wedge \bigwedge_{k+1 \leq h \leq n} (d\overrightarrow{x}, \overrightarrow{f}_h) = \det((\overrightarrow{e}_i, \overrightarrow{f}_h))\omega_1 \wedge \cdots \wedge \omega_q, \tag{2}$$

in (2), $\omega_1 \wedge \cdots \wedge \omega_q$ is the volume element of M^q.

The det $((\overrightarrow{e}_i, \overrightarrow{f}_h))$ is invariant under the choice of the frames $\overrightarrow{x}\,\overrightarrow{e}_1 \cdots \overrightarrow{e}_n$ and the choice or the frames $\overrightarrow{x}\,\overrightarrow{f}_1 \cdots \overrightarrow{f}_n$. Let us make this clear by the following consideration.

Take L^q to be the tangent space of M^q, then $\overrightarrow{e}_1 = \overrightarrow{f}_1, \ldots, \overrightarrow{e}_{k+q-n} = \overrightarrow{f}_{k+q-n}$, span L^{k+q-n}, $\overrightarrow{e}_{k+q-n+1}, \ldots, \overrightarrow{e}_q$ span the orthogonal complement of L^{k+q-n} in L^q, this is of dimension $q-(k+q-n) = n-k$. On the other hand the $(n-k)$ vectors $f_{k+1}, \ldots, f_n$ span the orthogonal complement $(L^k)^\perp$ of L^k in E. Since each of $\overrightarrow{e}_{k+q-n+1}, \ldots, \overrightarrow{e}_q$ must not lie in L^k, (otherwise it belongs to L^{k+q-n}), which is a contradiction, they must lie in the $(L^k)^\perp$, hence $f_{k+1}, \ldots, f_n$ and $\{\overrightarrow{e}_{k+q-n+1}, \ldots, \overrightarrow{e}_q\}$ are two bases of $(L^k)^\perp$. Let σ denote the identity linear transformation of $(L^k)^\perp \to (L^k)^\perp$, let i, h range over $k+q-n+1 \leq i \leq q, k+1 \leq h \leq n$, then

$$\sigma(e_i) = e_i \qquad \text{and} \qquad (\overrightarrow{e}_i, \overrightarrow{f}_h) = (\sigma(\overrightarrow{e}_i), \overrightarrow{f}_h).$$

This means that $\det((\overrightarrow{e}_i, \overrightarrow{f}_h))$ is just the determinant of the matrix of the linear transformation σ with reference to the base $(\overrightarrow{f}_h)$. But from the fundamental theory in linear algebra, we know that the determinant $|\sigma|$ is independent of the choice, of the base $(\overrightarrow{f}_h)$. A symmetric argument shows, that $|\sigma|$ is independent of the choice of $(\overrightarrow{e}_i)$.

But the choice of $(\overrightarrow{f}_h)$ and $(\overrightarrow{e}_i)$ is induced from the choice of the frames $\overrightarrow{x}\,\overrightarrow{e}_1, \ldots, \overrightarrow{e}_n$ and $\overrightarrow{x}\,\overrightarrow{f}_1, \ldots, \overrightarrow{f}_n$. This proves the invariance of the determinant.

This invariant det $((\overrightarrow{e}_i, \overrightarrow{f}_h))$ is one of the invariants of two linear subspaces in the space. There exist other invariants, for example, take $n = 3$, $k = 1$, $q = 2$.

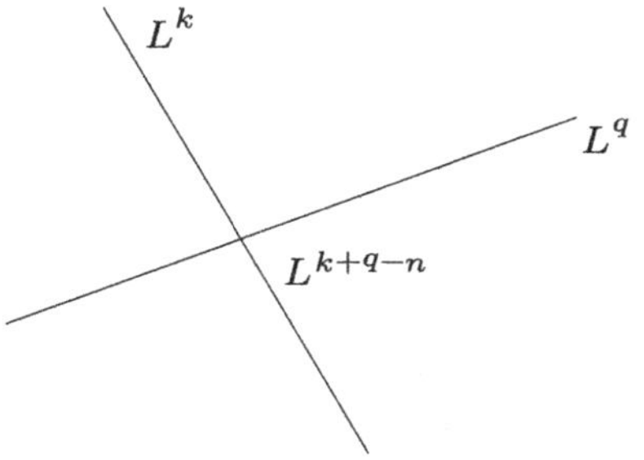

Fig. 46.

This is a plane and a line intersecting at a point, then the inclination angle of the line is also an invariant.

Multiplying (2) by the density of all L^k about a point, that is, the exterior product $\bigwedge_{\substack{1\le\alpha\le k\\ k+1\le s\le n}} \omega_{\alpha s}$ and then make, an integration.

On the left hand side, let us fix L^k first, then $\bigwedge_{k+1\le h\le n}(d\overrightarrow{x}, \overrightarrow{f}_h) = \bigwedge \omega_h$ and $\bigwedge(d\overrightarrow{x}, \overrightarrow{f}_h) = \bigwedge_{\substack{1\le\alpha\le k\\ k+1\le h,s\le n}} \omega_h\omega_{\alpha s} = \mathfrak{y}(n,k)$. The integration of Φ for a fixed L^k is just the $(k+q-n)$-dim volume of the intersection $L^k \cap M^q$, therefore

$$\text{L. H. S} = \int v_{k+q-n}(L^k \cap M^q)\mathfrak{y}(n,k).$$

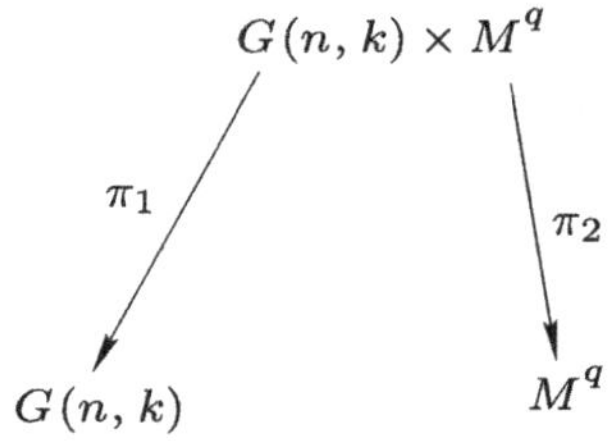

Fig. 47.

Note that if π_1, π_2 are projections, then fix L^k means that we make the projection π_1 first, and then integrate. We express this by saying that the integration process considered above is an integration with respect to the projection π_1. On the right hand side, we shall consider the integration with respect to the projection π_2, that is to say, we fix a point $\overrightarrow{x}$ in M^q and integrate first over all L^k about this point, the value $\int \det((\overrightarrow{e}_i, \overrightarrow{f}_h)) \wedge \omega_{\alpha s}$ is a universal constant C, since we have proved the invariance of det, this means that, to every L^k we have a det, it is the integration over all L^k about the fixed point $\overrightarrow{x}$, hence it only represents the measure of all L^k about the arbitrary fixed point $\overrightarrow{x}$, therefore it is independent of M^q.

Picking this constant out, and let $\overrightarrow{x}$ vary over all M^q, we have

$$\int \omega_q \wedge \cdots \wedge \omega_q = \nu(M^q).$$

Thus

$$\text{R. H. S.} = C\nu_q(M^q).$$

Integrations with respect to π_1 and π_2 are equivalent, hence

$$\int \nu_{k+q-n}(L^k \cap M^q)\mathfrak{y}(n,k) = C\nu_q(M^q) \tag{3}$$

Determination of the value of C

Of course we may calculate the value C directly from the integration of the det over all L^k about a fixed point, however this is rather complicated and difficult. It is important to make the following observation.

The same formula is valid for the spherical space with the same constancy: because C is the integration of $\det((\overrightarrow{e}_i, \overrightarrow{f}_h))$ over all L^k about a point, in the spherical space S^n, the unit sphere in $(n+1)$-dim space E^{n+1}, we introduce the same det, and find its C represents the same integration. Hence C is independent not only of the manifold M^q but also of the space considered, it is a universal constant.

In spherical space, $G(n,k)$ is the space of k-spheres on S^n. If we denote by $\widetilde{G}(n,k)$ the space of all oriented great k-spheres, and $G(n,k)$ the set of all great k-spheres (not oriented), then $\widetilde{G}(n,k)$ is a connected two-fold covering space of $G(n,k)$ which is locally homomorphic to $G(n,k)$.

Let M^q be a great sphere of dimension q, in spherical space S^n of curvature $+1$. Since L^k are all great k-spheres, we have $L^k \cap M^q = (k+q-n)$-dim. great sphere.

Therefore

$$\nu(L^k \cap M^q) = O_{k+q-n+1},$$
$$\nu(M^q) = O_{q+1},$$

according to our notation. Since the space is compact, we can integrate all over. From (3), we have for a spherical space S^n,

$$O_{k+q-n+1} \int \mathfrak{y}(n,k) = CO_{q+1}. \tag{4}$$

If $SO(m) =$ the proper (i.e. det $= 1$) orthogonal group in m variables, then the space of all oriented S^k can be identified with the rotation group from a fixed one S_0^k to any S^k. As has been explained in the beginning of the last lecture, we have

$$\widetilde{G}(n,k) = \frac{SO(n+1)}{SO(k+1) \times SO(n-k)}.$$

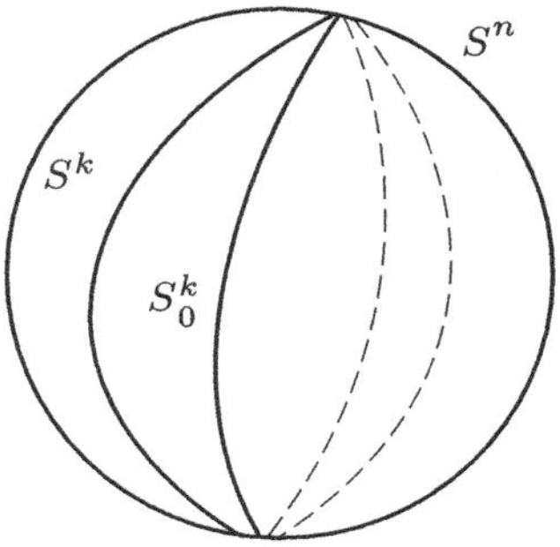

Fig. 48.

Hence we may assign the volume of $\widetilde{G}(n,k)$ to be

$$v(\widetilde{G}(n,k)) = \frac{\mathrm{vol}(SO(n+1))}{\mathrm{vol}(SO(k+1))vol(SO(n-k))}. \tag{5}$$

But the volume of the rotation group can be calculated from that of spheres.

On the oriented unit $(n-1)$-sphere S^{n-1}, the proper real orthogonal group is transitive. If $x_0 \in S^{n-1}$, the subgroup leaving x_0 fixed is just a proper orthogonal group $SO(n-1)$, since by the condition,

$$SO(n) = \{A : n \times n \text{ matrix} \mid AA^t = I, \det A = +1\},$$

the column vectors form an orthogonal system of n unit vectors.

The end points of these column vectors lie on the sphere S^{n-1}. Take $x_0 \in S^{n-1}$ to be the point with coordinates $(1,0,\ldots,0)$, then a matrix leaves x_0 fixed if and only if its first column vector has the form $\begin{pmatrix} 1 \\ 0 \\ \vdots \\ 0 \end{pmatrix}$. On account of its orthogonality, it takes the form $\begin{pmatrix} 1 & 0 & \ldots & 0 \\ 0 & & & \\ \vdots & & A_1 & \\ 0 & & & \end{pmatrix}$, that is we may identify it with an $(n-1) \times (n-1)$ proper orthogonal matrix. Hence we make the identification:

$$\text{oriented } S^{n-1} = SO(n)/SO(n-1).$$

We say that $SO(n)$ is a bundle over oriented S^{n-1} with fibre and group $SO(n-1)$.

Hence

$$\begin{aligned} v(SO(n)) &= v(S^{n-1})v(SO(n-1)) \\ &= O_n v(SO(n-1)) \\ &= O_n O_{n-1} \ldots O_3 v(SO(2)) \\ &= O_n O_{n-1} \ldots O_3 O_2, \end{aligned}$$

since $SO(2) \approx S^1$.

$$\int \mathfrak{y}(n,k) = v(G(n,k)) = \frac{1}{2} v(\widetilde{G}(n,k))$$

by (5)
$$= \frac{1}{2} \cdot \frac{O_{n+1} O_n \cdots O_2}{O_{k+1} \cdots O_2 O_{n-k} \cdots O_2}$$

$$= \frac{O_{n+1} O_n \cdots O_{n-k+1}}{O_{k+1} \cdots O_2 O_1} \qquad (\because O_1 = 2.)$$

By (4),

$$C = \frac{O_{k+q-n+1}}{O_{q+1}} \cdot \frac{O_{n+1} \cdots O_{n-k+1}}{O_{k+1} \cdots O_1}.$$

And we have proved the generalized Crofton's formula:

$$\int v_{k+q-n}(L^k \cap M^q) \mathfrak{y}(n,k) = \frac{O_{n+1} \cdots O_{n-k+1} O_{k+q-n+1}}{O_{k+1} \cdots O_1 O_{q+1}} v_q(M^q).$$

Applications

Consider 3-dimensional non-euclidean space X of constant curvature. Let T be a tetrahedron in X, let ℓ_i be the length of the sides ($i = 1, \dots, 6$), and α_i the dihedral angles at the sides, $\vee$ the volume of T.

We have no elementary simple formula relating these quantities. But we can prove that:

$$\int_{\text{plane } \cap T \neq \emptyset} \mathfrak{y} = \frac{1}{2} \left\{ \sum_{i=1}^{6} (\pi - \alpha_i) \ell_i + 2\varepsilon k \vee \right\}, \qquad \varepsilon = \pm 1, K > 0, \tag{6}$$

where $\mathfrak{y} = \mathfrak{y}(3,2)$ is the density of all planes in X, and εK is the curvature of X. If $\varepsilon = +1$, X is an elliptic space, if $\varepsilon = -1$, it is a hyperbolic space.

All planes intersecting T must intersect also the sides. Let v be the number of common points of a plane L^2 with the l-dimensional skeleton T^1 of T (that is to say, the polygon formed by the six edges of T). Considering $M^q = T^1$, and applying Crofton's formula, we have in this case $n = 3$, $k = 2$, $q = 1$ and $v = v(L^2 \cap T^1)$, and $v(T^1) = \sum_{i=1}^{6} \ell_i$, and $C = \frac{O_4 \not{O_3} \not{O_2} \not{O_1}}{\not{O_3} \not{O_2} \not{O_1} O_2} = \frac{2\pi^2}{2\pi} = \pi$, therefore

$$\int v \mathfrak{y} = \pi \sum_i \ell_i. \tag{7}$$

The general position of L^2 intersects with T^1 in 3 or 4 points. Let

$$\int_{v=3} \mathfrak{y} = F_3 \geqslant 0, \qquad \int_{v=4} \mathfrak{y} = F_4 \geqslant 0.$$

Hence (6) and (7) become

$$\begin{cases} F_3+F_4=\dfrac{1}{2}\left\{\displaystyle\sum_{i=1}^{6}(\pi-\alpha_i)\ell_i+2\varepsilon K\vee\right\}, \\ 3F_3+4F_4=\pi\displaystyle\sum_{i=1}^{6}\ell_i \end{cases}$$

or

$$\begin{cases} F_4=-\dfrac{1}{2}\pi\displaystyle\sum_{i=1}^{6}\ell_i+\dfrac{3}{2}\sum_i\alpha_i\ell_i-3\varepsilon K\vee\geq 0, \\ F_3=\pi\displaystyle\sum_i\ell_i-2\sum\alpha_i\ell_i+4\varepsilon k\vee\geq 0. \end{cases}$$

If X is an elliptic space $\varepsilon=+1$, if it is a hyperbolic space $\varepsilon=-1$, hence:

$$\begin{cases} \dfrac{1}{4}\displaystyle\sum_i(2\alpha_i-\pi)\ell_i\leq K\vee\leq\dfrac{1}{6}\sum_i(3\alpha_i-\pi)\ell_i, & \text{for elliptic space,} \\ \dfrac{1}{6}\displaystyle\sum_i(-3\alpha_i+\pi)\ell_i\leq K\vee\leq\dfrac{1}{4}\sum_i(\pi-2\alpha_i)\ell_i, & \text{for hyperbolic space.} \end{cases}$$

When K approaches to zero, $K\to 0$, we have the inequalities:

$$2\sum_i\alpha_i\ell_i-\pi\sum\ell_i\leq 0 \qquad \text{and} \qquad 3\sum\alpha_i\ell_i-\pi\sum\ell_i\geq 0,$$

or equivalently: $\frac{\pi}{3}\leq\frac{\sum\alpha_i\ell_i}{\sum\ell_i}\leq\frac{\pi}{2}$, for euclidean tetrahedron.

Moreover this is the best possible inequality we can obtain Further relationships between α_i, ℓ_i and $\vee$ can be found in the book of Polya-Szeg8, which is a famous exercise book. For example, there is a theorem asserting that, given six sides and six dihedral angles, then the tetrahedron T with these given ones as sides and angles is not uniquely determined. In the book, in fact, they prove this theorem by the method of integral geometry.

For an arbitrary polyhedron, we have $\sum_{\text{sides}}\ell_i\alpha_i$, which is a very important invariant (under motion). Minkowski discussed it in his geometric number theory, and introduced by it the Kantenkrümung (edge-curvature), which has a great significance.

At this step of development of integral geometry, we would like to look forward to the further progress in this: branch, and ask what are the most natural problems, that we may direct our efforts.

In my opinion I feel that the following problem is very fundamental: In a plane, we have L, A as $\cdots$ invariants, in 3-space we have L, A, v, $\cdots$ as invariants, I think we have to find out a complete system of invariants, and make a systematic research about the relationships among these invariants.

The principal method available for this systematic study is:

Consider a manifold $M^q \subseteq E^n$, where E^n may be euclidean or non-euclidean. Take the tube:

$$T_E(M^q) = \left\{ x \in E^n \mid \text{dist}(x, M^q) \leq \varepsilon \right\}.$$

Then make a calculation of the volume of this tube, if Z^n is a euclidean space, the formula of the volume is a polynomial of ε. And the coefficients of this polynomials are just the integral invariants containing L, A, V etc, which associate to M^q. Hence we can obtain a system of relations among invariants which satisfy many known formulas. As for the non-euclidean case, we no longer obtain a, polynomial of ε, in it some transcendental functions such as sine or cosine appear, although this complicates the treatment, we still can find systematically the relationships between the invariants.

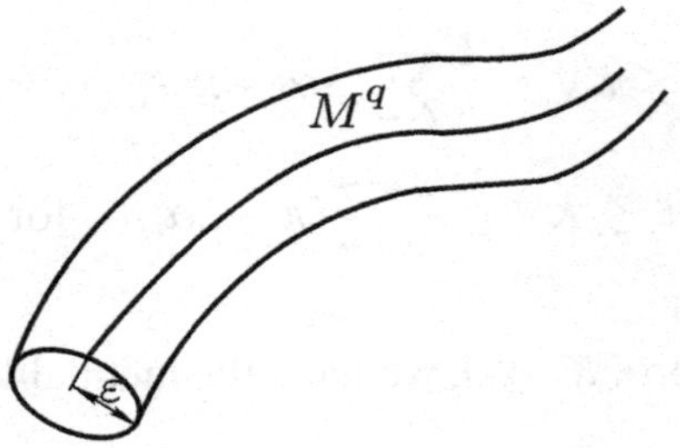

Fig. 49.

There is a paper concerning with this method.

H. Weyl: Volume of a tube (Amer. J. M. 1939. p.461–472).

The calculations of problems encountered in integral geometry of n dimensions are in general very complicated and difficult. Except for a few examples which have very strong invariant properties and very simple geometric meaning (such as the Crofton's formula discussed above), we do not have suitable and powerful instruments to solve them.

3.7 Lecture VII

Differential and integral invariants of a q-dimensional submanifold in a euclidean n-space

In fact we need not restrict to a euclidean space, and may consider the submanifold of a general homogeneous space of a Lie group. But it is better for us to confine ourselves in a euclidean space, since in this single case (as a matter of fact, it is not simple), we have already quite a lot of interesting results, and several unsolved problems.

Special cases:

A) $n = 2, q = 1$

Let γ be an oriented smooth curve having $\vec{x}(t)$ as its equation, therefore $(\vec{x}, \vec{x}) \neq 0$, let s be the arc length. (The smoothness assumption is not necessary, all the following considerations can be extended to a more general curve.)

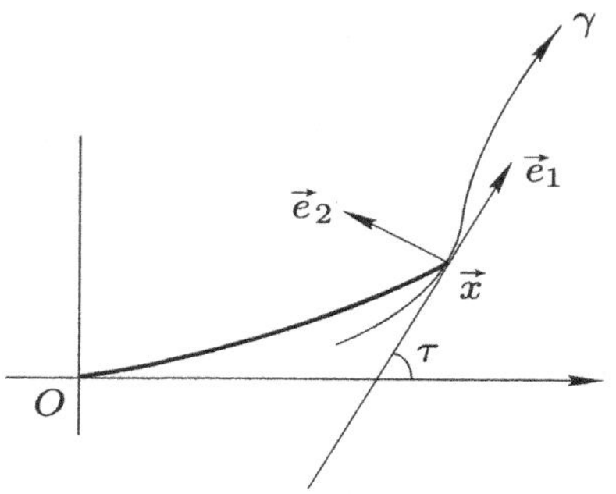

Fig. 50.

We want to find the invariants of the curve which are obtained by differentiation and integration. Associate a frame $\vec{x}\vec{e}_1\vec{e}_2$ to each point the curve γ, such that $\vec{e}_1$ is in the tangent direction, and $\vec{e}_2$ in the direction of the principal normal, hence $\vec{e}_i\vec{e}_j = \delta_{ij}$, $\det(\vec{e}_1, \vec{e}_2) = +1$, the orientation of the frame coincides with that of the plane. From the frenet formula

$$\frac{d\vec{e}_1}{ds} = k\vec{e}_2, \qquad \frac{d\vec{e}_2}{ds} = -k\vec{e}_1,$$

since, as proved in Lecture V, $d\vec{e}_i\vec{e}_i = 0$, $k \equiv \omega_{12} = -\omega_{21}$. The number k, being the coefficient of the infinitesimal displacement, is an invariant. If the value of k is small, then γ deviated only a little from a straight line, but if k is large, then γ becomes very curved. Hence k has simple geometric meaning, it measures the curvature of the curve, and is therefore called the curvature of the curve. We have found two differential invariants s, and k.

The integral invariants are obtained by integrate them among γ.

$$\int_0^L ds = L, \qquad \text{(this is the length of } \gamma)$$

and

$$\int_0^L kds.$$

The second integral invariants has an important geometric meaning. Observe that $\vec{e}_1 = (\cos\tau, \sin\tau)$, where τ is the angle of inclination of the tangent, then $\vec{e}_2 = (-\sin\tau, \cos\tau)$ is uniquely determined.

$$d\vec{e}_1 = (-\sin\tau d\tau, \cos\tau d\tau),$$

thus

$$kds = (d\vec{e}_1, \vec{e}_2) = (\sin^2\tau + \cos^2\tau)d\tau = d\tau.$$

We have $\int_0^L kds = \int_0^L d\tau =$ the angle that the tangent has rotated.

Especially, if γ is a closed curve, then the angle of rotation along the whole curve must be an integral multiple of 2π, This comes the formula

$$\frac{\int_\gamma kds}{2\pi} = \text{integer} \overset{\text{called}}{=\!=\!=} \text{rotation number of } \gamma.$$

Remark 3.7.1. *Although it seems rather obvious, the fact that the angle of rotation along a closed curve is $2n\pi$ requires a rigorous proof. The first clear and acceptable proof is given by H. Hopf in Composition Math. One can also find the proof in my lecture note. Selected topics in differential geometry, or perhaps in Kreuszing's book: Differential geometry.*

For example, in the 3 figures, each has rotation number $+1$, -1, 0.

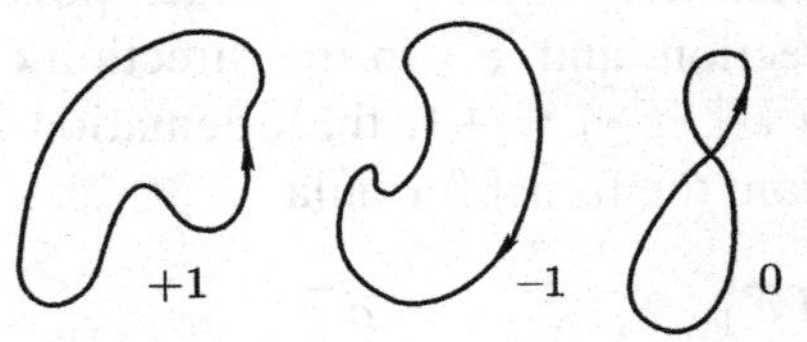

Fig. 51.

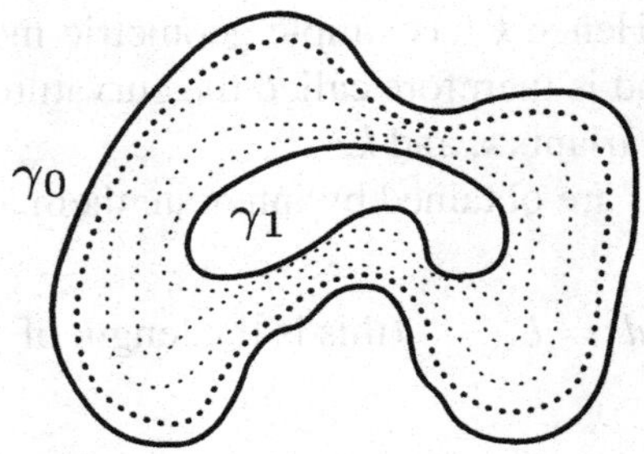

Fig. 52.

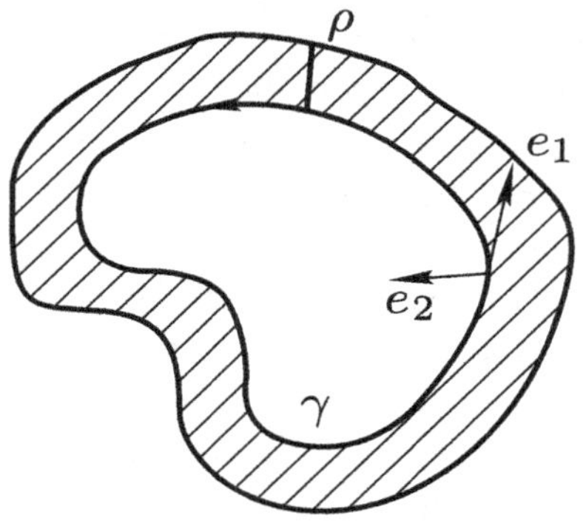

Fig. 53.

If two closed, smooth curves γ_0, γ_1, can be deformed to each other, and if in the process of deformation, each intermediate curve is closed and smooth, then we call this deformation a regular homotopy.

Theorem of H. Whitney: *Rotation number is invariant under regular homotopy.* (Some recent works of Smale have relation to this theorem).

Theorem on rotation number: *If γ is a simply closed curve (i.e. no self-intersection points), then the rotation number of γ is ± 1, where the sign is determined according to the orientation of γ and that of the plane.*

We shall show next that the above introduced integral invariants are very natural ones.

Let γ be a simply closed oriented smooth curve, take the tube T_ρ defined by

$$T_\rho = \{\vec{x} - r\vec{e}_2 \mid 0 \le r \le \rho\} \quad \text{i.e.}$$

a band of width f bounding outside of the curve γ.

Since $\vec{e}_1$, $\vec{e}_2$ is an orthonormal base, the area element is just the exterior product of the linear differential forms which are the coefficients of $d(\vec{x} - r\vec{e}_2)$ expressed as a linear combination of $\vec{e}_1$ and $\vec{e}_2$.

$$\begin{aligned}\frac{d\vec{x}}{ds} &= \vec{e}_1, d(\vec{x} - r\vec{e}_2)\\ &= ds\vec{e}_1 - dr\vec{e}_2 + rkds\vec{e}_1\\ &= (1+rk)ds\vec{e}_1 - dr\vec{e}_2,\end{aligned}$$

therefore the area element of the domain T_ρ is $-(1+kr)ds \wedge dr = (1+kr)dr \wedge ds$.

$$\begin{aligned}V(T_\rho) &= \int_0^L \int_0^\rho (1+kr)drds = \int_0^L \left(\rho + \frac{1}{2}k\rho^2\right) ds\\ &= \rho L + \frac{\rho^2}{2}\int_0^L kds = \rho L + \frac{\rho^2}{2}\cdot 2\pi = \rho L + \pi\rho^2.\end{aligned}$$

This is the Steiner's formula which we have encountered already.

$V(T_\rho)$ is a polynomial in ρ, with coefficients L, π or more exactly, L, $2\pi(=\int_0^L kds)$ which are the integral invariants considered above.

We can therefore state that: for a suitable closed curve, when we consider the area of the tube T_ρ associated with it, then the integral invariants appears naturally as the coefficients of the polynomial representing $V(T_\rho)$.

In differential geometry, the assumption that a curve is smooth is not advisable, because most of the curves do not satisfy this assumption. In any way, the above formula $V(T_\rho) = \rho L + \rho^2\pi$ can be generalized to a piecewise smooth closed curve. For example, let us consider a triangle, then T_ρ consists of A_1, A_2, A_3 and B_1, B_2, B_3. $V(A_1 \cup A_2 \cup A_3) = L\rho$, while B_1, B_2, B_3 constitute a circle with ρ as radius, hence $V(B_1 \cup B_2 \cup B_3) = \pi\rho^2$.

Therefore $V(T_\rho) = \rho L + \rho^2\pi$.

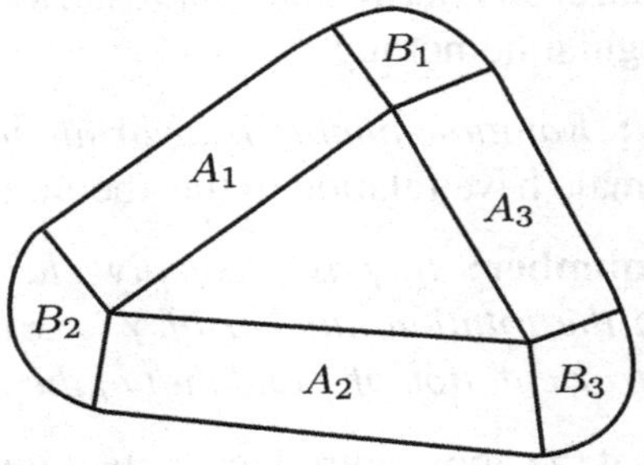

Fig. 54.

B) Next we consider the case when $n = 3, q = 2$

Consider an oriented surface, since it is oriented, a unique unit normal $\vec{e}_3$ is determined at each point $\vec{x}$ on the surface. We can assign a frame $\vec{x}\,\vec{e}_1\,\vec{e}_2\,\vec{e}_3$ to the point $\vec{x}$, therefore $\vec{e}_1$, $\vec{e}_2$ are tangent vectors at $\vec{x}$.

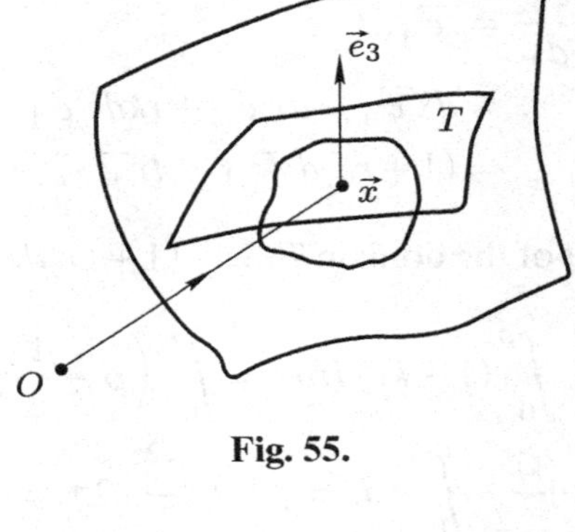

Fig. 55.

Define $\mathrm{I} = (d\vec{x}, d\vec{x})$ = the first fundamental from, (note that this inner product is symmetric).
$\mathrm{II} = (d\vec{x}, d\vec{e}_3)$ = the second fundamental form.
$\mathrm{III} = (d\vec{e}_3, d\vec{e}_3)$ = the third fundamental form.

It is a positive definite quadratic form, it is an inner property of the surface, and has no relation to the space in which the surface is imbedded.

If (u, v) are local coordinates in a neighborhood of $\vec{x}$, then $\vec{x} = x(u, v)$,

$$
\begin{aligned}
d\vec{x} &= x_u du + x_v dv, \\
(d\vec{x}, d\vec{x}) &= (x_u \cdot x_u)du^2 + 2(x_u \cdot x_v)dudv + (x_v \cdot x_v)dv^2 \\
&= Edu^2 + 2Fdudv + Gdv^2.
\end{aligned}
$$

This is a positive definite form, i.e. $EG - F^2 > 0$, $E > 0$.

On the other hand II contains $d\vec{e}_3$, hence it has a relationship with the space in which the surface is imbedded. We can prove that in general III is a linear combination of I, and II, by this reason, III does not give new informations, it is sufficient, to consider III only. Although I and II are invariant (quadratic differential forms), in our seeking of the differential invariants, we do not satisfied with these I and II, since they are not functions on the surface. Only when we are given a local coordinates (u, v) that we can determine the functions E, F, G, hence they can not be regarded as functions of the surface itself. A function of the surface must be independent of local coordinates. Our immediate aim is to get the invariant functions.

Consider I $+t$II where t is a parameter. Take the ratio, $\frac{\det(\mathrm{I}+t\mathrm{II})}{\det(\mathrm{I})}$, the numerator and enumerator are both 2×2 determinants, hence this ratio can be expanded as a polynomial in t of the second degree in the form

$$1 + 2Ht + Kt^2$$

since when $t = 0$, the ratio is 1 this is just the constant term. When local coordinates (u, v) are given

$$
\begin{aligned}
\mathrm{I} &= (x_u \cdot x_u)du^2 + 2(x_u \cdot x_v)dudv + (x_v \cdot x_v)dv^2 \\
&= Edu^2 + 2Fdudv + Gdv^2,
\end{aligned}
$$

det (I) is defined to be $\begin{vmatrix} (x_u\, x_u) & (x_u\, x_v) \\ (x_v\, x_u) & (x_v\, x_v) \end{vmatrix} = \begin{vmatrix} E & F \\ F & G \end{vmatrix}$, det (II) is similarly defined.

It can be shown that the quotient $\frac{\det(\mathrm{I}+t\mathrm{II})}{\det(\mathrm{I})}$ is independent of the choice of local coordinates. This owes to the fact that in evaluating the determinant of a quadratic differential form, the effect of a coordinate transformation only amounts to multiplying a factor which has relation to the Jacobian of the transformation. But we are considering the ratio of two determinants, thus this factor must be canceled out.

$1 + 2Ht + Kt^2$ is an invariant, because I, II are invariants. Moreover, as has been shown, it is independent of the local coordinates. Therefore H and K are scalar functions on the surface.

Usually H and K are called the mean curvature and the Gaussian curvature respectively of the surface, They are the most important differential invariants of the surface. If the surface is compact, we may integrate over it.

Let Φ be the area element of the surface, then $\int H\Phi$ and $\int K\Phi$ are integral invariants. They can be found naturally as the coefficients of the polynomial representing the volume of the tube associated to this surface. But this is contained in the following more general discussions.

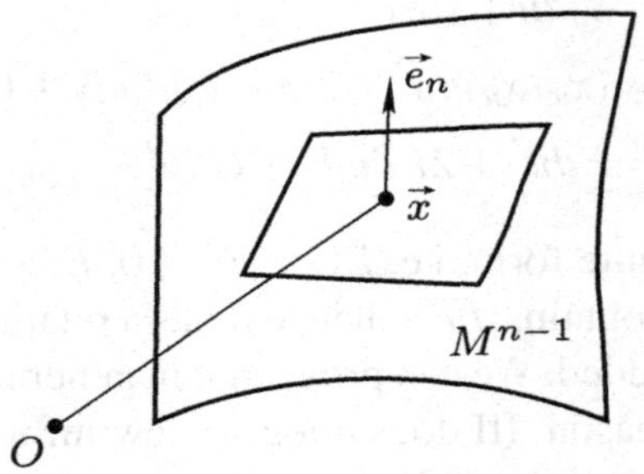

Fig. 56.

C) $q = n-1$

Let M^{n-1} be an oriented hypersurface in n-space. It has a uniquely determined unit normal vector $\overrightarrow{e}_n$.

Define in a similar manner

I $= (d\overrightarrow{x}, d\overrightarrow{x}) > 0$, which is a +def form.
II $= (d\overrightarrow{x}, den)$.

Take I $+t$ II and consider the ratio $W = \frac{\det(\mathrm{I}+t\mathrm{II})}{\det(\mathrm{I})}$, both of the determinants are $(n-1)\times(n-1)$, hence $W = \sum_{k=0}^{n-1}\binom{n-1}{k}\sigma_k t^k$ where $\binom{n-1}{k}$ are binomial coefficients. As have been explained, W is independent of the local coordinates, furthermore it is an invariant, therefore the coefficients $\sigma(=1), \sigma_1, \ldots, \sigma_{n-1}$ are scalar invariants of M^{n-1}.

Anyone who is familiar with differential geometry knows that on the surface if r_1, r_2 are the principal curvatures, then $H = \frac{r_1+r_2}{2}$, $K = r_1 r_2$. Similarly, it can be proved that the roots of $W = 0$ are principal curvatures, hence $\sigma_1, \ldots, \sigma_{n-1}$ are the elementary symmetric functions of principal curvatures.

Let Φ be the volume element of M^{n-1}, if we consider the local coordinates $(u^1, \ldots, u^{n-1})$, then it is well known that $\Phi = \sqrt{g}du^1 \wedge \cdots \wedge du^{n-1}$, where $g = \det(I)$.

If M^{n-1} is compact, we have $\mu_i \stackrel{\mathrm{def}}{=\!=} \int_{M^{n-1}} \sigma_i \Phi$ as the integral invariants $0 \le i \le n-1$.

We have obtained n integral invariants of the hypersurface M^{n-1} among them μ_0 represents the volume of M^{n-1}. These integral invariants occur naturally as the coefficients of the polynomial representing the volume of the tube associated to M^{n-1}. More clearly:

$$T_\rho = \{\vec{x} + r\vec{e}_n \mid 0 \le r \le \rho\}.$$

Take frames $\vec{x}\vec{e}_1, \dots, \vec{e}_n$, so that $\vec{e}_1, \dots, \vec{e}_{n-1}$ are tangent vectors at $\vec{x}$, $d\vec{x}$, being a tangent vector, is a linear combination of $\vec{e}_1, \dots, \vec{e}_{n-1}$,

$$d\vec{x} = \omega_1 \vec{e}_1 + \cdots + \omega_{n-1}\vec{e}_{n-1}.$$

If $d\vec{e}_n = \theta_1 \vec{e}_1 + \cdots + \theta_{n-1}\vec{e}_{n-1} + \theta_n \vec{e}_n$, from the fact that

$$0 = \frac{1}{2} d(\vec{e}_n, \vec{e}_n) = (\vec{e}_n, d\vec{e}_n) = \theta_n, \ \theta_n \text{ must be equal to zero.}$$

Since $\vec{e}_n$ is uniquely determined, the linear differential forms θ_i are linear combinations of the differential forms $du_1, \dots, du_{n-1}$, where $(u_1, \dots, u_{n-1})$ is any local coordinate system. Also ω_j are linear combinations of du_α. But ω_j are independent, hence θ_i are linear combination of ω_j,

$$\theta_\alpha \overset{\text{def}}{=\!=} \sum_\beta h_{\alpha\beta}\omega_\beta, \qquad 1 \le \alpha, \beta \le n-1.$$

We shall prove first that $h_{\alpha\beta} = h_{\beta\alpha}$.

$$0 = d(d\vec{x}) \equiv - \sum_{1 \le \alpha \le n-1} \omega_\alpha \wedge d\vec{e}_\alpha, \ \text{mod}\ \vec{e}_1, \dots, \vec{e}_{n-1}.$$

From $(\vec{e}_\alpha, \vec{e}_n) = 0$, we have $0 = (d\vec{e}_\alpha, \vec{e}_n) + (\vec{e}_\alpha, d\vec{e}_n) = (d\vec{e}_\alpha, e_n) - \theta_\alpha$ thus

$$d\vec{e}_\alpha \equiv -\theta_\alpha \vec{e}_n, \ \text{mod}\ \vec{e}_1, \dots, \vec{e}_{n-1},$$

hence

$$0 = \sum_{\alpha=1}^{n-1} \omega_\alpha \wedge de_\alpha = \sum_\alpha (\omega_\alpha \wedge \theta_\alpha)\vec{e}_n + \ell_{-c} \text{ of } \vec{e}_1, \dots, \vec{e}_{n-1}.$$

Since $\vec{e}_1, \dots, \vec{e}_n$ are linearly independent,

$$\sum_\alpha \omega_\alpha \wedge \theta_\alpha = 0 \to \sum_{\alpha,\beta} h_{\alpha\beta}\omega_\alpha \wedge \omega_\beta = 0,$$

we conclude that $h_{\alpha\beta} = h_{\beta\alpha}$, in fact, e.g. the coefficient of $\omega_1 \wedge \omega_2$ is $h_{12} - h_{21} = 0$, thus $h_{12} = h_{21}$, etc. In general, we have a Cartan's lemma which assures that:

If $\sum h_{\alpha\beta} x_\alpha x_\beta = 0$, then $h_{\alpha\beta} = -h_{\beta\alpha}$.
If $\sum h_{\alpha\beta}\omega_\alpha \wedge \omega_\beta = 0$, then $h_{\alpha\beta} = h_{\beta\alpha}$.
$\mathrm{I} = (d\vec{x}, d\vec{x}) = \sum_\alpha \omega_\alpha^2$ (symmetric multiplication).

$\mathrm{II} = (d\vec{e}_n, d\vec{x}) = \sum_\alpha \omega_\alpha \theta_\alpha = \sum_{\alpha\beta} h_{\alpha\beta}\omega_\alpha\omega_\beta$ (symmetric multiplication).

Since $W = \frac{\det(\mathrm{I}+t\mathrm{II})}{\det(\mathrm{I})}$ is independent of local coordinates $(u_1, \dots, u_{n-1})$, i.e. independent of the choice of the base $du_1, \dots, du_{n-1}$ of linear differential forms, we may take $\omega_1, \dots, \omega_{n-1}$ as a basis.

Therefore,

$$\det I = \begin{vmatrix} 1 & & & 0 \\ & 1 & & \\ & & \ddots & \\ 0 & & & 1 \end{vmatrix} = 1$$

$$W = \det(\mathrm{I}+t\mathrm{II}) = \det(\delta_{\alpha\beta} + th_{\alpha\beta}) = \sum_{\alpha=1}^{n-1} \binom{n-1}{k} \sigma_k t^k,$$

$$\begin{aligned} d(\vec{x} + r\vec{e}_n) &= d\vec{x} + rd\vec{e}_n + dr\vec{e}_n \\ &= \sum_\alpha \omega_\alpha \vec{e}_\alpha + r\sum_\alpha \theta_\alpha \vec{e}_\alpha + dr\vec{e}_n \\ &= \sum_{\alpha,\beta} \delta_{\alpha\beta}\omega_\beta \vec{e}_\alpha + r\sum_{\alpha\beta} h_{\alpha\beta}\omega_\beta \vec{e}_\alpha + dr\vec{e}_n \\ &= \sum_\alpha \left[\sum_\beta (\delta_{\alpha\beta} + rh_{\alpha\beta})\omega\right] \vec{e}_\alpha + dr\vec{e}_n. \end{aligned}$$

Since $\{\vec{e}_i\}$ is an orthonormal base, the volume element of the tube T_ρ is the exterior product of the coefficients in the above formula, which is:

$$\begin{aligned} \bigwedge_\alpha \left(\sum_\beta (\delta_{\alpha\beta} + rh_{\alpha\beta})\omega_\beta\right) \wedge dr &= \det(\delta_{\alpha\beta} + rh_{\alpha\beta})\omega_1 \wedge \cdots \wedge \omega_{n-1} \wedge dr \\ &= \sum_{k=0}^{n-1} \binom{n-1}{k} r^k \sigma_k \Phi \wedge dr. \end{aligned}$$

By integration, we have if M^{n-1} is compact

$$\begin{aligned} V(T_\rho) &= \sum_{k=0}^{n-1} \binom{n-1}{k} \int_{M^{n-1}} \sigma_k \Phi \int_0^\rho r^k dr = \sum_{k=0}^{n-1} \binom{n-1}{k} \frac{1}{k+1} \mu_k \rho^{k+1} \\ &= \sum_{k=0}^{n-1} \frac{(n-1)!}{(k+1)!(n-k-1)!} \cdot \frac{n}{n} \cdot \mu_k \rho^{k+1} \\ &= \sum_{i=1}^{n} \frac{n!}{i!(n-i)!} \cdot \frac{1}{n} \cdot \mu_{i-1} \rho^i \\ &= \frac{1}{n} \sum_{i=1}^{n} \binom{n}{i} \mu_{i-1} \rho^i . \end{aligned}$$

This formula for the volume of the tube is polynomial in ρ, and the set of integral invariants of the hypersurface appear naturally as the coefficients of this polynomial.

Taking $n = 2$, we have $V(T_\rho) = \mu_0 \rho + \frac{1}{2} \mu_1 \rho^2$,
$n = 3$, we have $V(T_\rho) = \mu_0 \rho + \mu_1 \rho^2 + \frac{1}{3} \mu_2 \rho^3$.

Except μ_0, which represents the volume of M^{n-1} the last one μ_{n-1} has a simple but important topological significance. It can be proved that, if M^{n-1} has no boundary, then μ_{n-1} represents the degree of the normal mapping.

Taking S_0^{n-1} to be the unit sphere in n-space, the end point of $\overrightarrow{e}_n$ must therefore fall upon S_0^{n-1}. To each point $\overrightarrow{x}$ in M^{n-1} we assign its normal vector $\overrightarrow{e}_n$, this constitutes a map from $M^{n-1} \to S_0^{n-1}$, and is called the Gauss normal mapping. If M^{n-1} is closed, we may count how many times the image of M^{n-1} covers S_0^{n-1} this is called the Brouwer's degree of the map. This is a topological property, only depends on the topology of M^{n-1}.

If $(n-1)$ is even we have moreover

$$\mu_{n-1} = \frac{1}{2} \chi(M^{n-1}),$$

where $\chi(M^{n-1})$ is the Euler-Poincaré characteristic.

μ_i are all metric invariants, and we have pointed out that μ_{n-1} is moreover a topological invariant. The problem as to construct invariants of M^{n-1} under more general group, such as topological invariants, are something very interesting in differential topology.

3.8 Lecture VIII

To a surface in the three space, we obtained σ_1, σ_2 as differential invariants, and μ_0 (= the area of the surface), μ_1, μ_2 as integral invariants. If the surface is closed and has no boundary, $\mu_2 = \int \sigma_2 \Phi =$ the integral of Gaussian curvature, is the degree of the Gauss normal mapping and is equal to a constant times the Euler characteristic. This is, in fact, the Gauss-Bonnet theorem. Moreover

$$\mu_1 = \int \sigma_1 \Phi = \text{integral of mean curvature.}$$

This μ_1 is a very important integral invariant.

The above discussions can be extended to sectionally smooth surfaces. If the surface used is a convex polyhedron, then μ_1 is called the Kantenkrümmung (edge-curvature) of Minkowski and has value $\sum_i \alpha_i \ell_i$. This Kantenkrümmung plays a fundamental role in the treatments of convex body.

μ_0, μ_1, μ_2 can be defined as invariants of a very general species of domains.

Definition 3.8.1. Let D be domains in E^3, a function, defined over domains D is completely additive, if $\varphi(D+D') = \varphi(D) + \varphi(D') - \varphi(D \cap D')$, φ is invariant under motion if $\varphi(T(D)) = \varphi(D)$, where T represents a motion.

According to this definition, μ_0, μ_1, μ_2 and V (volume) are completely additive and are invariant under motion. Hence any linear combination of them' is also completely additive and invariant under motion.

It is unsolved whether there exist other completely additive and invariant functions except those generated, from μ_0, μ_1, μ_2 and V. Or to what kind of family of subsets of E^3 can we define on it completely additive and invariant functions.

But if we restrict to certain special family of subsets, it has been proved that on this special family μ_0, μ_1, μ_2, V are the only completely additive and invariant functions.

q-dimensional submanifold in E^n

The differential invariants of M^q is less obvious in this case, hence it is better to perform the methods of tube used above, and obtain the integral invariants first. Consider the tube

$$T_\rho = \{x \mid \text{dist}\,(x, M) \leq \rho\}.$$

If M^q in E^n is compact and has no boundary point, then H. Weyl had proved a formula (cf. Volume of a tube. Amer. J. H. 1939. p. 461–472).

$$V(T_\rho) = O_m \sum_{0 \leq e \leq q} \mu_e \frac{\rho^{m+e}}{m(m+2)\cdots(m+e)},$$

where $m = n - q$ is the codimension of M^q, and e is even, O_m is the volume of a unit sphere in E^m, i.e. the $(m-1)$-dim area.

This is a polynomial having terms $\rho^{n-q}, \rho^{n-q+2}, \rho^{n-q+4}, \ldots$.

It does not include the case of a hypersurface, i.e. we can not obtain the formula for M^{n-1} by putting $q = n-1$. The reason lies in the fact that: If we consider the unit normal bundle. In the case of M^{n-1}, each fibre consists of two (disconnected) points, say, A and B. But in the case of M^q, for example M^1 in M^3, as indicated in the forgoing figure, a fibre is no longer disconnected, rather it is connected. It is due to this connectedness that we encounter much more difficulties in calculations.

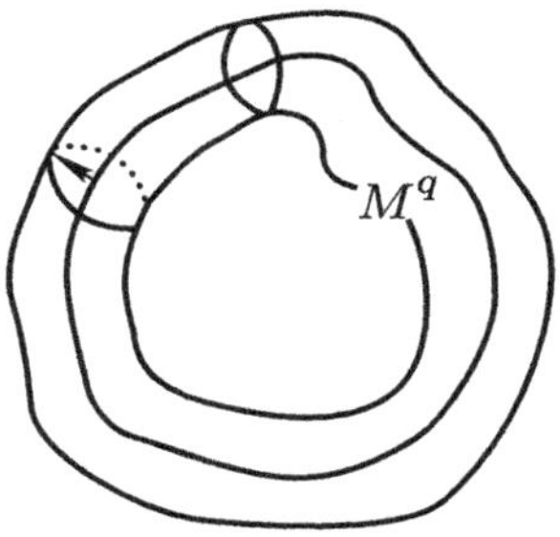

Fig. 57.

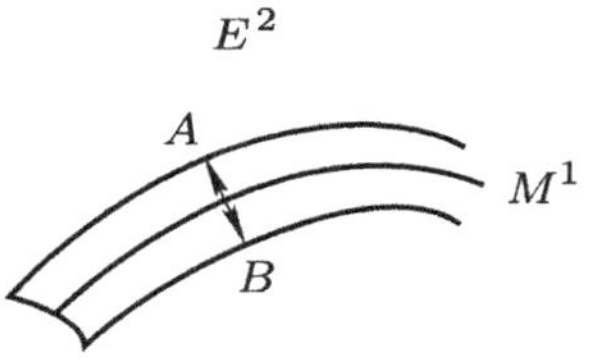

Fig. 58.

The Weyl's formula is therefore less accurate than that in the case for a hypersurface. In fact, if in the case of M^{n-1}, we calculate the volume of both

$$T_1(\rho) = \{\overrightarrow{x} + r\overrightarrow{e}_n \mid 0 \leq r \leq \rho\} \quad \text{and} \quad T_2(\rho) = \{\overrightarrow{x} - r\overrightarrow{e}_n \mid 0 \leq r \leq \rho\}$$

and add together, the terms containing μ_{odd} cancel each other, only μ_e remain, and we obtain the Weyl's formula for M^{n-1}.

μ_e are the principal integral invariants of M^q.

If Φ represents the volume element of M, and define $\mu_e = \int \sigma_e \Phi$, then σ_e are differential invariants of M^q. It can be proved that they take the form

$$\sigma_e = \text{const} \sum_{i,j} \delta\begin{pmatrix} i_1, \ldots, i_e \\ j_1, \ldots, j_e \end{pmatrix} R_{i_1 i_2 j_1 j_2}, \ldots, R_{i_{e-1} i_e i_{e-1} i_e},$$

in which $R_{ijk\ell}$ are the components of the Riemann-Christoffel tensor (relative to orthogonal frames) of the induced Riemannian metric on M^q, and $\delta\binom{i_1,\ldots,i_e}{j_1,\ldots,j_e}$ does not vanish if only if $j_1, \ldots, j_e$ are rearrangement of $i_1, \ldots, i_e$, its value is $+1$ if the permutation is even, and -1 if odd.

The Riemann-Christoffel tensor R is not a scalar invariant, it is a tensor, but the σ_e which are obtained from the Riemann tensor are scalar invariants, in fact σ_e are among the most important scalar invariants of the Riemann metric.

For example $\sigma_2 = \text{const} \sum R_{i_1 i_2 j_1 j_2}$ is called the scalar curvature of the Riemann metric, hence σ_e may be regarded, as generalized scalar curvature.

If q is even, and $e = q$, then σ_e is just the integral which appears in the Gauss-Bonnet formula.

If $q+k \geq n$, one, can prove a formula of the Crofton's type

$$\int \mu_e(L^k \cap M^q)\mathfrak{y}(n,k) = \text{const } \mu_e(M^q).$$

This Lie product property may be considered as a generalization of Crofton's formula, since when $t=0$, μ_e represents volume, i.e. $\mu_0(L^k \cap M^q) = v_{k+q-n}(L^k \cap M^q)$, etc.

For a hypersurface, we have also

$$\int \mu_i(L^k \cap M^{n-1})\mathfrak{y}(n,k) = \frac{O_{n-1},\ldots,O_{n-k+1}}{O_1 O_2,\ldots,O_{k-1}} \frac{O_{n-i+1}}{O_{k-i+1}} \mu_i(M^{n-1}).$$

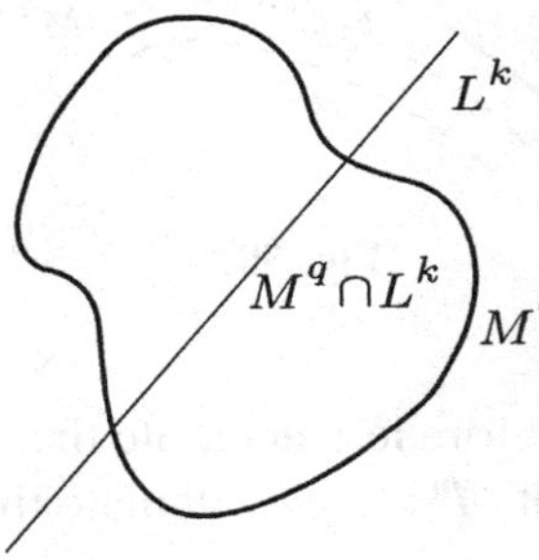

Fig. 59.

Let us recite without proof several generalizations of kinematic formulas.

For M^{n-1} we have some good results, but for M^q we have only a few loose results.

I. If M_0 and M_1 are two hypersurfaces, M_0 is fixed, M_1 is moving, and $\partial D_0 = M_0$, $\partial D_1 = M_1$, then

$$O_n \int \chi(D_0 \cap D_1)\mathscr{K}(M_1)$$
$$= O_2,\ldots,O_n \left\{ O_n(\chi_0 V_1 + \chi_1 V_0) + \frac{1}{n}\sum_{h=0}^{n-2}\binom{n}{h+1}\mu_h(M_0)\mu_{n-2-h}(M_1) \right\}.$$

where V_i are volumes of D_i, χ_i are Euler characteristics of D_i (cf. Amer. J. M. 1952).

This formula holds also in the non-euclidean case, as was proved by Santalo. In the proof we need the first main theorem of algebraic invariants and other tools.

II. M^q: fixed, M^r: moving, then

$$\int u_e(M^q \cap M^r)\mathscr{K}(M^r) = \sum_{j:\text{even}} C_j \mu_j(M^q)\mu_{e-j}(M^r)$$

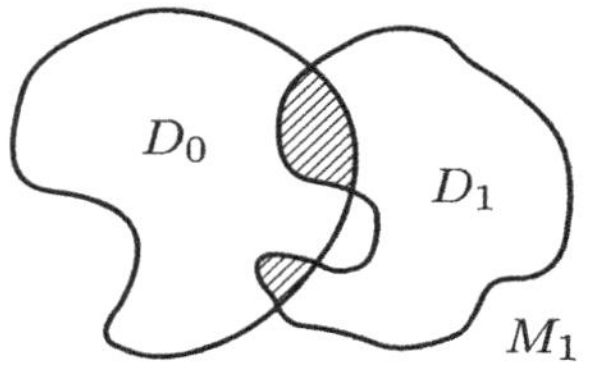

Fig. 60.

where C_j are certain universal constants. I think that this formula must hold also in the non-euclidean, case.

Integral geometry in arbitrary homogeneous space

Up to the present we consider only the invariants under the group of motions, and we confine ourselves to the case of euclidean spaces. This policy is taken because we have to make our object clear in this special case; however, we encounter already many difficulties in this euclidean case.

We shall now discuss briefly the integral geometry in a general homogeneous space, and obtain the invariants under more general groups of transformations.

Take G to be a Lie group, H a closed subgroup.

G/H = the space of all left cosets (a left coset is a set $gH = \{gh|h \in H\}$, for a fixed $g \in G$).

G/H can be topologized to become a topological space. And we can define the action of G/H by G in the following way:

$$G \times G/H \to G/H,$$
$$(g_1, g^H) \to (g_1 g^H).$$

From the property of left cosets, it is easily seen that this actually defines an action of G/H by G, hence G may be considered as the group of transformations of the topological space G/H.

Example 3.8.2. G: the group of all motions in E^n. H: the subgroup of G leaving the origin O fixed. Let T_x denote the translation carrying O to the point $\overrightarrow{x}$. If a motion S carries O to $\overrightarrow{x}$, then $T_x^{-1}S$ leaves O fixed, that is, $T_x^{-1}S \in H$, or $S \in T_x H$.

Conversely any motion belonging to $T_x H$ carries O to $\overrightarrow{x}$. This proves the important fact:

The point of a euclidean space can be considered as a coset of motions $E^n \approx G/H$. The closed subgroup H, which leaves one point fixed, is called the group of isotropy. Therefore, a euclidean space is a very special case of homogeneous space. The discussions of integral geometry can be translated to discussions of a general homogeneous space, by taking G to be any arbitrary Lie group.

Example 3.8.3. Let $G = SL(n,R)$, the special linear group = the group of all $n \times n$ matrices with real elements and det $= +1$.

H be the set of all matrices in G, having the form

$$\begin{pmatrix} 1 & \\ 0 & \\ \vdots & * \\ 0 & \end{pmatrix},$$

H is a subgroup, since $\begin{pmatrix} 1 & \\ 0 & \\ \vdots & * \\ 0 & \end{pmatrix}\begin{pmatrix} 1 & \\ 0 & \\ \vdots & * \\ 0 & \end{pmatrix} = \begin{pmatrix} 1 & \\ 0 & \\ \vdots & * \\ 0 & \end{pmatrix}$, moreover, the inverse of $\begin{pmatrix} 1 \; a_{12}\cdots a_{1n} \\ 0 \\ \vdots \quad * \\ 0 \end{pmatrix}$ is $\begin{pmatrix} A_{11}\Delta^{-1} & \\ A_{12}\Delta^{-1} & * \\ A_{1n}\Delta^{-1} & \end{pmatrix} = \begin{pmatrix} 1 & \\ 0 & \\ \vdots & * \\ 0 & \end{pmatrix}$ since $|A_{11}| \cdot 1 = 1$, and $|A_{12}| = \cdots = |A_{1n}| = 0$, $\Delta = 1$.

Consider the G/H, when do $g_1, g_2 \in G$ belong to the same left coset relative to H?

$\begin{pmatrix} g_{11}\cdots g_{1n} \\ \vdots \quad \vdots \\ g_{n1}\cdots g_{nn} \end{pmatrix}\begin{pmatrix} 1 & \\ 0 & \\ \vdots & * \\ 0 & \end{pmatrix} = \begin{pmatrix} g_{11} & \\ \vdots & * \\ g_{n1} & \end{pmatrix}$, hence all matrices in the same coset have the same first column. Conversely, if two matrices have the same first column. Since $SL(n,R)$ is a group, we may suppose that

$$\begin{pmatrix} g_{11} & g_{12} & \cdots & g_{1n} \\ \vdots & \vdots & & \vdots \\ g_{n1} & g_{n2} & \cdots & g_{nn} \end{pmatrix}\begin{pmatrix} h_{11} & \cdots & h_{1n} \\ \vdots & & \vdots \\ h_{n1} & \cdots & h_{nn} \end{pmatrix} = \begin{pmatrix} g'_{11} & g'_{12} & \cdots & g'_{1n} \\ \vdots & \vdots & & \vdots \\ g'_{n1} & g'_{n2} & \cdots & g'_{nn} \end{pmatrix}.$$

By the assumption

$$\begin{cases} g_{11}h_{11} + \cdots + g_{1n}h_{n1} = g'_{11} = g_{11}, \\ \cdots\cdots\cdots\cdots \\ g_{n1}h_{11} + \cdots + g_{nn}h_{n1} = g'_{n1} = g_{n1}, \end{cases} \quad \text{i.e.} \begin{cases} g_{11}(h_{11}-1) + \cdots + g_{1n}h_{n1} = 0, \\ \cdots\cdots\cdots\cdots \\ g_{n1}(h_{11}-1) + \cdots + g_{nn}h_{n1} = 0. \end{cases}$$

Since the Δ of this system of equations is not zero, we have only the trivial solution, i.e. $h_{11} - 1 = h_{21} = \cdots = h_{n1} = 0$. This proves that two matrices have the same first column is a necessary and sufficient condition for their belonging to the same coset.

The first column vector is necessary not equal to zero, since the matrix is non-singular. Hence we conclude the important result:

$G/H \approx V^n \setminus 0\}$, where V^n is the space of first column vectors. If G/H and G/K are two homogeneous spaces with the same G, the elements of which are g^H and g'^K respectively. We want to define the conception of the “incidence” of g^H and g'^K.

Definition 3.8.4. g^H and $g'K$ are incident iff they have a non-empty intersection.

It is easily proved that this definition include the ordinary incident relations. The special case when one of the two groups G or K is discrete, we have an important application of integral geometry to the geometric number theory. Let us explain it more clearly. Suppose $G = SL(n,R)$, K is the subgroup of G whose elements are all integers, (we shall denote the set of all integers by Z), we shall prove that G/K is the space of all lattices.

Consider any element of gK.

$$\begin{pmatrix} g_{11} & \cdots & g_{1n} \\ \vdots & & \vdots \\ g_{n1} & \cdots & g_{nn} \end{pmatrix} \begin{pmatrix} k_{11} & \cdots & k_{1n} \\ \vdots & & \vdots \\ k_{n1} & \cdots & k_{nn} \end{pmatrix} = \begin{pmatrix} (g_{11}k_{11} + \ldots + g_{1n}k_{n1}), \ldots \\ \\ (g_{n1}k_{11} + \ldots + g_{nn}k_{n1}), \ldots \end{pmatrix}, k_{ij} \in Z.$$

From this formula, we find that: The column vectors of the product are linear combinations, with integer coefficients, of the column vectors of (g_{ij}).

For example, take $n = 2$, draw the two column vectors (g_{11}, g_{21}) and (g_{12}, g_{22}). Then the column vectors of the coset g^K belong to the lattice generated by the column vectors of (g_{ik}). And K is the transformation, subgroup which leaves this lattice invariant.

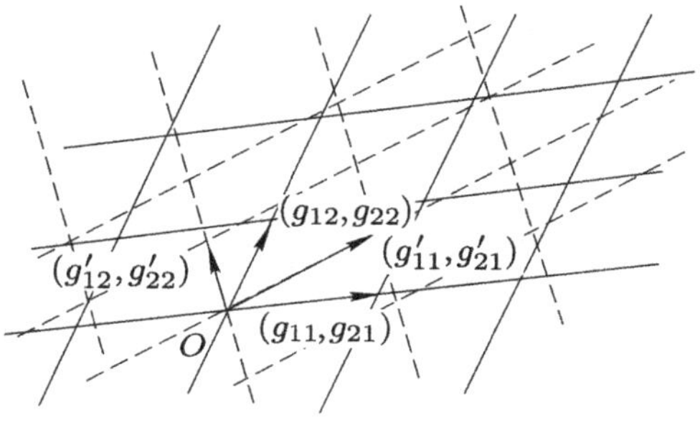

Fig. 61.

In the figure, we draw by dotted lines another lattice which is generated by another element (g'_{ik}) of G. Hence a coset g^K corresponds to a lattice, and the homogeneous space just corresponds to the space of all lattices. As was stated above.

Definition 3.8.5. A lattice point $P: \begin{pmatrix} x_1 \\ \vdots \\ x_n \end{pmatrix}$ (in a lattice) is called primitive, if $\{\text{g.c.d.}(x_1, \ldots, vx_n)\} = 1$ that is, OP does not contain other lattice points. Take H to be the subgroup considered in Example 3.8.3, then $G/H \approx V^n - \{O\}$.

It can be proved that an element $x \in G/H$ is incident with an element $g^K = \lambda$ of G/K iff x is a primitive lattice point of λ.

According to this result, the above definition of incidency has a very natural geometric meaning. Let us show how we can apply the integral geometry (considered for the homogeneous spaces G/K, G/H) to the geometric number theory. Minkowski has proved a fundamental theorem in geometric number theory. Consider a convex domain D in E^n containing the origin O, which is symmetric with respect to O.

Theorem 3.8.6. *If $v(D) > 2^n$, then D contains non-zero lattice point. (In fact, from $G/H \approx V \backslash \{O\}$. O can not be regarded as a lattice point, it should be cast out from the consideration.) This theorem is very important, since it shows the existence, of the solution of many Diophantine equations.*

Conversely we would like to ask: How good is this estimate? More exactly: How small the domain must be, that there exists a lattice such that in D we have no (non-zero) primitive lattice point.

Although Minkowski was not able to solve this problem, he payed much attentions to it, because he stated many times his conjecture about this problem. This problem remained unsolved for almost fifty years.

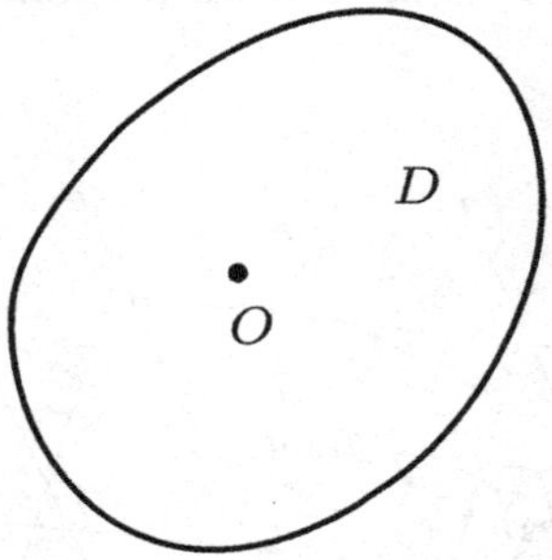

Fig. 62.

E. Hlawka – Siegel theorem

If $v(D) < \zeta(n)$, then $\exists$ a lattice of det $= 1$ which has no non-zero lattice point in D.

In the theorem $\zeta(n)$ represents the Riemann Zeta function. Of course we must have $\zeta(n) < 2^n$.

E. Hlawka solved it by a purely geometric number theoretic method, but Siegel applied the integral geometry in solving it.

In fact, this theorem is a consequence of Crofton's formula.

If $D^* = D \backslash \{O\}$, then $D^* \subseteq G/H$.

To each lattice λ of G/K (which is the space of all lattices), take all the primitive lattice points of λ belonging to D^*, its number is denoted by N.

Moreover, let $\mathscr{L}_n$ = density in G/K, and the given domain D^* is not necessary convex, but $v(D^*) < +\infty$. Then we have a formula of Crofton's type:

$$\int N\mathscr{L}_n = \nu(D^*)\zeta(2)\cdots\zeta(n-1),$$

where the integration is over all lattices. From this formula we can prove Hlawka's theorem. As for the case $n = 3$ of this theorem, we may confer the book of Santalo.

The concept of integral geometry is extended from the euclidean space and the group of motion to a general homogeneous space of a Lie group.

By this extension we enlarge the domain of integral geometry to a very far extent, and integral geometry can be applied more, effectively to various branch of mathematics by this extension. But we must observe that the homogeneous space generally does not have a G-invariant measure, e,g. the n-dim affine space under the action of $GL(n/R) = G$, does not have a G-invariant measure. Only when we restrict to $SL(n,R)$ can we obtain a measure. The problem concerning the existence of a G-invariant measure is completely solved. For general topological groups, cf. A. Weil: L'intération dans les groups topologiques et ses applications. For Lie groups cf. Santalo. Especially for connected Lie groups, it can be solved in terms of structure constants.

Recent approaches

Radon transform: (cf. Helgason. Bulletin A. M. S. 1964?, this is a general survey in this field, and there can be found a complete list of bibliography.)

Let $E = E^2$ be the plane. $f : E \to \mathbb{R}$ be a smooth real valued function with compact support, i.e. the value of f vanishes outsider compact domain D. Consider any straight line L in the plane. Along L, we may consider the integration

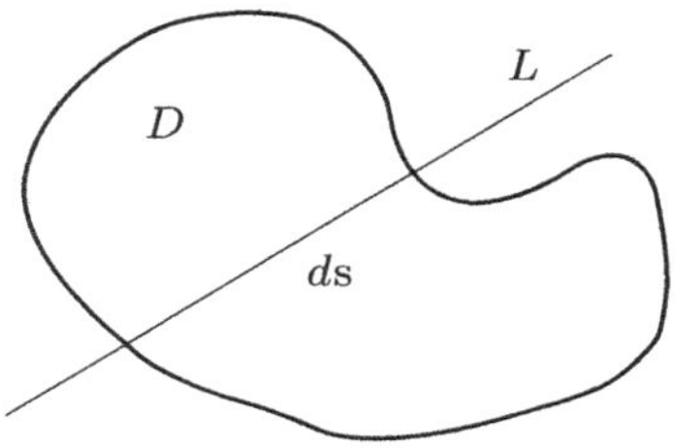

Fig. 63.

$$\hat{f}(L) = \int_L f(x)ds \in \mathbb{R}.$$

If E^* denotes the space of line, then we have

$$\hat{f} : E^* \to \mathbb{R}.$$

Conversely we may ask, if $\hat{f}$ is given, can we determine f uniquely? This is called the Radon's problem.

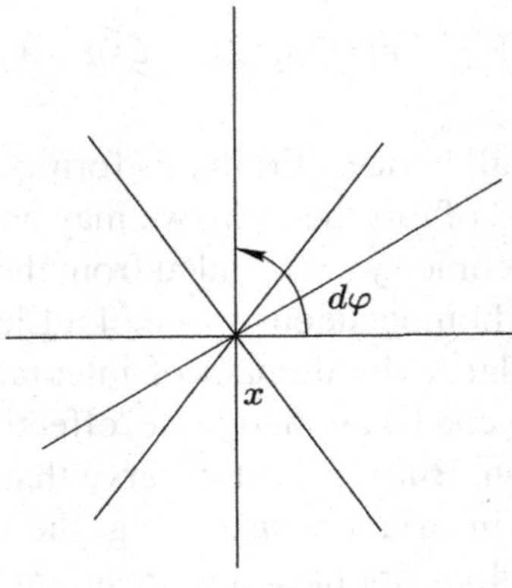

Fig. 64.

We may also formulate the problem in an alternate way. Given a function $g : E^* \longrightarrow \mathbb{R}$, fix an arbitrary point x, and consider the set of all straight lines through x. Then the function $\breve{g}: E \longrightarrow \mathbb{R}$ defined by $\breve{g}(x) = \int_{\text{all lines through } x} g d\varphi$ is also a Radon transform:

Another question is: from f we have $\hat{f}$, from $\hat{f}$ we have $(\hat{f})^V$, then what is the relationship between f and $(\hat{f})^V$.

In general, if in a homogeneous space we are given two families of spaces, then we can make a Radon transform from the functions defined on one family to that defined on the other family. By this process we can sometimes simplify the problems at hand.

Futz John has applied this method in solving the initial values problems of wave functions.

Gelfand have much work in the recent integral geometry. He applied successfully the integral, geometry to the representation problem of a non-compact Lie group. One can find his papers (translated into English) in the collected papers of Amer. M. S.

Part III Differentiable Manifolds

4

Multilinear Algebra

4.1 The tensor (or Kronecker) product

We recall that a vector space V over a field F is a set of elements in which two operations are defined:

1) addition of two elements of V giving as sum an element of V;
2) multiplication of an element of V by an element of F (to the left), giving an element of V.

These operations are moreover assumed to satisfy the conditions:

I. Relative to addition V is an abelian group.
II. The second operation has the properties:
   1) $\alpha(x+y) = \alpha x + \alpha y$,
   2) $(\alpha+\beta)x = \alpha x + \beta x$, $\alpha, \beta \in F$, $x, y \in V$,
   3) $\alpha(\beta x) = (\alpha\beta)x$,
   4) $0 \cdot x = 0$, $1 \cdot x = x$.

In 4) the symbols $0, 1$ in the left-hand sides are of F, while the O in the right-hand side is the unit element of V as an abelian group.

The elements of V are called *vectors* and the elements of F *scalars*. We shall assume throughout our discussions that F has characteristic zero. With few exceptions we also assume that V is finite-dimensional. The latter means, as is well-known, that there are n vectors $x_1, \ldots, x_n$, such that any vector can be expressed uniquely as their linear combination, with coefficients in F.

A function f in V, with values in F, is called linear, if

$$f(\alpha x + \beta y) = \alpha f(x) + \beta f(y) \tag{1}$$

for all vectors x, y and scalars α, β. We define the addition of two linear functions f, g, and scalar multiplication of a linear function according to the equations

$$
\begin{aligned}
(f+g)(x) &= f(x)+g(x), \\
(\alpha f)(x) &= \alpha f(x).
\end{aligned}
\tag{2}
$$

It is then easily verified that the set of all linear functions in V forms a vector space, called the dual or conjugate space of V and to be denoted by V^*.

An element $x \in V$ gives rise to a scalar-valued linear function in V^*, taking the value $x^*(x)$ for $x^* \in V^*$. It can be proved that every linear function in V^* can be defined this way, so that V can be identified with the space of linear functions in V^*. To express this "duality" relation, more clearly, we introduce the notation

$$\langle x, x^* \rangle = x^*(x). \tag{3}$$

The function $\langle x, x^* \rangle$ is then defined in the Cartesian product $V \times V^*$ taking scalar values, and is "bilinear":

$$
\begin{aligned}
\langle \alpha x + \beta y, x^* \rangle &= \alpha \langle x, x^* \rangle + \beta \langle y, x^* \rangle, \\
\langle x, \alpha x^* + \beta y^* \rangle &= \alpha \langle x, x^* \rangle + \beta \langle x, y^* \rangle.
\end{aligned}
\tag{4}
$$

To a base x_i, $i = 1, \ldots, n$, in V there is a uniquely defined "dual base", x_k^*, $k = 1, \ldots, n$, satisfying the conditions

$$\langle x_i, x_k^* \rangle = \delta_{ik}, \tag{5}$$

where δ_{ik} is the Kronecker index taking the value 1 when $i = k$ and the value 0 otherwise.

We shall now study the relations between several vector spaces. When more than one vector space is under consideration, they are understood to be over the same field, and the convention will be adopted that reference to the field be dropped.

A map $f : V \to Z$ of a vector space V into a vector space Z is called *linear*, if

$$f(\alpha x + \beta y) = \alpha f(x) + \beta f(y) \tag{6}$$

for all vectors $x, y \in V$ and scalars α, β. This notion generalizes a linear function. By defining the addition of two linear maps $f, g : V \to Z$ and the scalar multiplication of a linear map by the equations analogous to (2):

$$
\begin{aligned}
(f+g)(x) &= f(x)+g(x), \\
(\alpha f)(x) &= \alpha f(x),
\end{aligned}
\tag{7}
$$

we see that all the linear maps of V into Z form a vector space, to be denoted by $\mathscr{L}(V;Z)$.

Definition 4.1.1. If V, W, Z are vector spaces, a map $f : V \times W \to Z$ is called *bilinear*, if

$$
\begin{aligned}
f(\alpha_1 x_1 + \alpha_2 x_2, y) &= \alpha_1 f(x_1, y) + \alpha_2 f(x_2, y), \\
f(x, \alpha_1 y_1 + \alpha_2 y_2) &= \alpha_1 f(x, y_1) + \alpha_2 f(x, y_2),
\end{aligned}
\tag{8}
$$

for any α_1, $\alpha_2 \in F$, $x, x_1, x_2 \in V$, $y, y_1, y_2 \in W$. Similarly, we define a *multilinear map*: $V_1 \times \cdots \times V_s \to Z$.

With a definition of addition and scalar multiplication similar to the case of linear maps, all bilinear maps $V \times W \longrightarrow Z$ can be made into a vector space $\mathscr{L}(V,W;Z)$. Moreover, we verify easily the isomorphisms

$$\mathscr{L}(V_1,W;Z) \cong \mathscr{L}(V;\mathscr{L}(W,Z)) \cong \mathscr{L}(W;\mathscr{L}(V;Z)). \tag{9}$$

We wish to reduce the study of bilinear maps of $V \times W$ into that of linear maps by solving the problems: Given V, W, to find a vector space Y and a bilinear map $h: V \times W \longrightarrow Y$, depending only on V, W, such that to any bilinear map $f: V \times W \longrightarrow Z$ into any vector space Z there exists a linear map $g: Y \longrightarrow Z$ satisfying $f = gh$. Schematically this means that commutativity holds in the diagram:

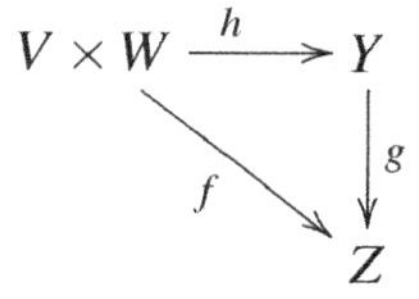

(10)

We proceed to prove that Y exists and is unique up to an isomorphism, provided that Y is spanned "minimal" in the sense that it is spanned by the image $h(V \times W)$.

Theorem 4.1.2. *Let V, W be vector spaces, and let (h_i, Y_i), for each $i = 1,2$, consist of vector space Y_i and a bilinear map $h_i: V \times W \longrightarrow Y_i$ such that:*

1) *The image $h_i(V \times W)$ spans Y_i;*
2) *To any vector space Z and any bilinear map $f: V \times W \longrightarrow Z$ there exist linear maps $g_i: Y_i \longrightarrow Z$ satisfying $f = g_i h_i$.*

Then there exists a one-one isomorphism $u: Y_1 \longrightarrow Y_2$ such that $h_2 = uh_1$.

To prove the theorem we apply property 2) of (h_1, Y_1) to the bilinear map: $h_2: V \times W \longrightarrow Y_2$. It follows that there is a linear map $\varphi_1: Y_1 \longrightarrow Y_2$, with $h_2 = \varphi_1 h_1$. Similarly, there is a linear map $\varphi_2: Y_2 \longrightarrow Y_1$, satisfying $h_1 = \varphi_2 h_2$. Combining these two relations, we get $h_1 = \varphi_2 \varphi_1 h_1$ and $h_2 = \varphi_1 \varphi_2 h_2$. Since $h_i(V \times W)$ spans Y_i, it follow that $\varphi_2 \varphi_1$ is the identity isomorphism of Y_1 onto itself and $\varphi_1 \varphi_2$ is the identity isomorphism of Y_2 onto itself.

The theorem is proved by taking $u = \varphi_1$.

We shall now construct a pair (h, Y) with the desired properties. Let $V \circ W$ be the linear space which consists of all formal sums

$$\sum \alpha_{xy}(x,y), \qquad \alpha_{xy} \in F,$$

where $\alpha_{xy} = 0$ except for a finite number of pairs (x,y) in $V \times W$. Let S be the linear subspace of $V \circ W$ generated by all elements of the form

$$(\alpha_1 x_1 + \alpha_2 x_2, \beta_1 y_1 + \beta_2 y_2) = \sum_{i,j=1}^{2} \alpha_i \beta_j (x_i, y_j), \qquad \alpha_i, \beta_j \in F.$$

Here we adopt the convention to identify (x,y) with $1(x,y)$, where by $V \times W$ is imbedded as a subset of $V \circ W$. Denote the factor space $V \times W/S$ by $V \otimes W$, and let $h : V \times W \to V \otimes W$ be the natural projection which assigns to (x,y) the coset to which it belongs. We shall prove that his bilinear and that the pair $(h, V \otimes W)$ has the properties 1), 2), of Theorem 4.1.2.

By definition the element

$$(\alpha_1 x_1 + \alpha_2 x_2, y) - \alpha_1(x_1, y) - \alpha_2(x_2, y)$$

belongs to S. Hence its image under h is zero, or,

$$h(\alpha_1 x_1 + \alpha_2 x_2, y) = \alpha_1 h(x_1, y) + \alpha_2 h(x_2, y).$$

This, and a similar relation, prove that h is bilinear. Property 1) is valid by definition. To verify 2) let $f : V \times W \to Z$ be given, define a linear map $\overline{f} : V \circ W \to Z$ by

$$\overline{f}(\sum \alpha_{xy}(x,y)) = \sum \alpha_{xy} f(x,y).$$

Then $\overline{f}$ maps S to zero and induces a linear map $g : V \otimes W \to Z$. For this g we have

$$gh(x,y) = g((x,y) + S) = \overline{f}((x,y)) = f(x,y),$$

for any $(x,y) \in V \times W$.

We shall also prove that $V \otimes W$ is finite-dimensional assuming of course that both V and W are. Let $e_1, \ldots, e_m$ be a base for V and $f_1, \ldots, f_n$ be a base for W. Then we have

$$h\left(\sum_{i=1}^{m} \alpha_i e_i, \sum_{j=1}^{n} \beta_j f_j\right) = \sum_{i,j} \alpha_i \beta_j h(e_i, f_j).$$

Hence any element $h(x,y)$ is a linear combination of $h(e_i, f_j)$, which are finite in number. Since $h(V \times W)$ spans $V \otimes W$, the $h(e_i, f_j)$ span $V \otimes W$, which is therefore finite-dimensional and, in fact, is of dimension $\leq mn$.

Definition 4.1.3. The pair $(h, V \otimes W)$ is called the tensor or Kronecker product of the vector spaces V and W.

We introduce the notation

$$x \otimes y = h(x,y), \qquad x \in V, \qquad y \in W. \tag{11}$$

Quite frequently we will call the space $V \otimes W$ the tensor product of V and W without mentioning the mapping h.

Theorem 4.1.4. *There is an isomorphism between the space of bilinear maps $V \times W \to Z$ and the space of linear maps $V \otimes W \to W$:*

$$\mathscr{L}(V,W;Z) \cong \mathscr{L}(V \otimes W;Z). \tag{12}$$

To a linear map $g : V \otimes W \to Z$ the map $f = gh : V \times W \to Z$ is bilinear. This defines a map

$$\mathscr{L}(V \otimes W; Z) \to \mathscr{L}(V, W; Z),$$

which is clearly linear. It is onto, because to a given f there exists g with $f = gh$. Moreover, $gh = 0$ implies, since $h(V \times W)$ spans $V \otimes W$, that $g = 0$. Hence the kernel is zero,

Corollary 4.1.5. *$(V \otimes W)^*$ is isomorphic to the space of bilinear maps of $V \times W$ into F.*

Henceforth we will consider these spaces to be identified by means of this isomorphism.

Let $e_1, \cdots, e_m; f_1, \ldots, f_n$ be bases of V and W respectively. Define the bilinear map $\varphi_{ij} : V \times W \to F$ by $\varphi_{ij}(e_k, f_\ell) = \delta_{ik}\delta_{j\ell}$, $i, k = 1, \ldots, m$; $j, \ell = 1, \ldots, n$. By bilinearity φ_{ij} is defined for any element of $V \times W$. These bilinear maps are linearly independent in $\mathscr{L}(V, W; F)$. For it

$$\sum_{i,j} \alpha_{ij}\varphi_{ij} = 0, \qquad \alpha_{ij} \in F,$$

we have

$$0 = \sum_{i,j} \alpha_{ij}\varphi_{ij}(e_k, f_\ell) = \sum_{i,j} \alpha_{ij}\delta_{ik}\delta_{j\ell} = \alpha_{k\ell}.$$

Hence the vector space $\mathscr{L}(V, W; F)$ or, by Theorem 4.1.4 the vector space $(V \otimes W)^*$ is of dimension $\geq mn$. But the latter has the same dimension as $V \otimes W$, which is $\leq mn$. It follows that the dimension of $V \otimes W$, is equal to mn. We state this result in the theorems:

Theorem 4.1.6. *If V is dimension m and W is of dimension n, then $V \otimes W$ is of dimension mn. If $e_1, \ldots, e_m$ and $f_1, \ldots, f_n$ are bases of V and W respectively, then $e_i \otimes f_j$, $i = 1, \ldots, m$, $j = 1, \ldots, n$, form a base of $V \otimes W$.*

This property of dimension characterizes in a sense the tensor product, as given by the theorem:

Theorem 4.1.7. *Let V, W be vector spaces of dimensions m, n respectively. Let $\varphi : V \times W \to Y$ be a bilinear map such that $\varphi(V \times W)$ spans Y and that Y has the dimension mn. Then Y is isomorphic to $V \otimes W$.*

By the main property of a tensor product, these exists a linear map $g : V \otimes W \to Y$ such that $\psi = hg$. For the base vectors e_i, f_j of V, W we have then

$$\psi(e_i, f_j) = gh(e_i, f_j) = g(e_i \otimes f_j).$$

Since Y is of dimension mn and $\psi(V \times W)$ spans Y, the vectors $\psi(e_i, f_j)$ are linearly independent. The same is therefore true of the vectors $g(e_i \otimes f_j)$. Since $e_i \otimes f_j$ form a base of $V \otimes W$, the theorem follows.

The isomorphism between Y and $V \otimes W$ is natural, in a sense which we do not intend to make precise. We say that Y and $V \otimes W$ are *naturally* or *canonically isomorphic.*

Corollary 4.1.8. $(V \otimes W)^*$ *is canonically isomorphic to* $V^* \otimes W^*$.

To prove this we define the map $\psi : V^* \times W^* \to (V \otimes W)^*$ by the relation.

$$\langle x \otimes y, \psi(x', y') \rangle = \langle x, x' \rangle \langle y, y' \rangle$$

for any $x \in V$, $x' \in V^*$, $y \in W$, $y' \in W^*$. This map is clearly well defined, and is bilinear. If e'_k and f'_ℓ are bases V^* and W^* dual to e_i and f_j respectively, so that

$$\langle e_i, e'_k \rangle = \delta_{ik}, \ \langle f_j, f'_\ell \rangle = \delta_{j\ell},$$

we have

$$\langle e_i \otimes f_j, \psi(e'_k, f'_\ell) \rangle = \delta_{ik} \delta_{j\ell}.$$

This proves that $\psi(e'_k, f'_\ell)$ form a dual base of $e_i \otimes f_j$, and hence span $(V \otimes W)^*$. The corollary thus follows from the theorem.

Corollary 4.1.9. *The space of endomorphisms of* V *is canonically isomorphic to* $V \otimes V^*$.

We recall that the space of endomorphisms of V is by definition the vector space $\mathscr{L}(V;V)$. We define a map $\psi : V \times V^* \to \mathscr{L}(V;V)$ by the condition

$$\psi(x, x')(y) = \langle y, x' \rangle x, \tag{13}$$

for any $x, y \in V$, $x' \in V^*$. This map is clearly bilinear. If e_i, e'_k, $i, k = 1, \ldots, n$, are dual bases in V and V^* respectively, we have

$$\psi(e_i, e'_k)(e_j) = \langle e_j, e'_k \rangle e_i = \delta_{jk} e_i = \begin{cases} e_i, \ j = k, \\ 0, \ j \neq k \end{cases} \quad j = 1, \ldots, n.$$

This proves that $\psi(V \times V^*)$ spans $\mathscr{L}(V;V)$. Since the latter is of dimension n^2, the corollary is proved.

We observe from the proof that $\psi_{ik} = \psi(e_i, e'_k)$ form a base of $\mathscr{L}(V;V)$ and that the isomorphism between $V \otimes V^*$ and $\mathscr{L}(V;V)$ is established by mapping $e_i \otimes e'_k$ into ψ_{ik}. On the other hand, we can define a natural bilinear map $f : V \times V^* \to F$ by

$$f(x, x') = \langle x, x' \rangle, \quad x \in V, x' \in V^*.$$

To this there corresponds a linear map $g : V \otimes V^* \to F$ defined by

$$g(x \otimes x') = \langle x, x' \rangle.$$

Identification of $V \otimes V^*$ and $\mathscr{L}(V;V)$ gives a linear map $t : \mathscr{L}(V;V) \to F$, which assigns a scalar to every endomorphism of V. This scalar is called the *trace* of the endomorphism. In fact, if, in terms of the above base, the endomorphism is $\sum_{i,k} \alpha_{ik} \psi_{ik}$, $\alpha_{ik} \in F$, we have, by definition,

$$t\left(\sum_{i,k} \alpha_{ik} \psi_{ik}\right) = g\left(\sum_{i,k} \alpha_{ik}(e_i \otimes e'_k)\right) = \sum_{i,k} \alpha_{ik} \langle e_i, e'_k \rangle = \sum_i \alpha_{ii}.$$

We state without the proof the following theorems on tensor products:

Theorem 4.1.10 (Commutativity of tensor products). *The tensor products $V \otimes W$ and $W \otimes V$ are naturally isomorphic.*

Theorem 4.1.11 (Associativity of tensor products). *The tensor products $V \otimes (W \otimes Z)$ and $(V \otimes W) \otimes Z$ are naturally isomorphic.*

Definition 4.1.12. Let $f : V \longrightarrow W$, $f' : V' \longrightarrow W'$ be linear maps. To them there is associated a bilinear map $g : V \times V' \longrightarrow W \otimes W'$ defined by $g(x,x') = f(x) \otimes f'(x')$, $x \in V, x' \in V'$. The map g defines in turn a linear map $\varphi : V \otimes V' \longrightarrow W' \otimes W'$ such that

$$\varphi(x \otimes x') = g(x,x') = f(x) \otimes f'(x').$$

The map φ is called the tensor product of the linear maps f and f', and will by denoted by $f \otimes f'$.

From this definition follows immediately the theorem:

Theorem 4.1.13. *Let*

$$V \xrightarrow{f} W \xrightarrow{g} Z,$$

$$V' \xrightarrow{f'} W' \xrightarrow{g'} Z'$$

be vector spaces connected by the linear maps indicated. Then

$$(g \circ f) \otimes (g' \circ f') = (g \otimes g') \circ (f \otimes f').$$

In particular, if f, f' are one-one isomorphisms with g, g' as their inverses, then $f \otimes f'$ is a one-one isomorphism of $V \otimes V'$ into $W \otimes W'$ with $g \otimes g'$ as its inverse.

The above consideration can be extended to the case of any finite number of vector spaces. We state the result without proof:

Theorem 4.1.14. *To the Cartesian product $V_1 \times \cdots \times V_s$ of vector spaces there exists a pair $(h, V_1 \otimes \cdots \otimes V_s)$ consisting of a vector space $V_1 \otimes \cdots \otimes V_s$, and a multilinear map $h : V_1 \times \cdots \times V_s \longrightarrow V_1 \otimes \cdots \otimes V_s$ such that to any multilinear map $f : V_1 \times \cdots \times V_s \longrightarrow Z$ there is a linear map $g : V_1 \otimes \cdots \otimes V_s \longrightarrow Z$ satisfying $f = gh$. The space $V_1 \otimes \cdots \otimes V_s$ is determined up to an isomorphism by the further condition that it is spanned by $h(V_1 \times \cdots \times V_s)$. Moreover, it is naturally isomorphic to the spaces $((V_1 \otimes V_2) \otimes V_3 \cdots) \otimes V_s$, etc.*

4.2 Tensor spaces

Definition 4.2.1. Let V be a vector space and V^* its dual space. An element of the tensor product

$$V_s^r = \underbrace{V \otimes \cdots \otimes V}_{r \text{ times}} \otimes \underbrace{V^* \otimes \cdots \otimes V^*}_{s \text{ times}}$$

is called a *tensor of type* (r,s), r being the *contravariant order* and s the *contravariant order*. In particular, an element of V_0^r is called a *covariant tensor of order r and an element of* V_s^0 *a covariant tensor of order s*. By convention V_0^0 will mean the field of scalars. A tensor of order one is also called a vector, so that the elements of V will be called *contravariant vectors* and those of V^* *covariant vectors*.

Remark 4.2.2. *In practice the* V*- and* V^**-factors in* V_s^r *may alternate. We arrange them in the above order merely for convenience.*

In the following we shall adopt the convention to drop the multiplication sign in a tensor product when there is no confusion. Moreover, all indices will run from 1 to n, the dimension of V, unless otherwise specified.

Let e_i, e'^k be dual bases in V and V^* so that

$$\langle e_i, e'^k \rangle = \delta_i^k. \tag{14}$$

These bases give to a base in V_s^r consisting of the elements

$$e_{i1} \cdots e_{ir} e'^{k_1} \cdots e'^{ks}. \tag{15}$$

A tensor $x \in V_s^r$ can then be written in a unique way in the form

$$x = \sum_{(i)(k)} \xi_{k_1 \cdots k_s}^{i_1 \cdots i_r} e_{i_1} \cdots e_{i_r} e'^{k_1} \cdots e'^{k_s}, \tag{16}$$

with the components $\xi_{k_1 \ldots k_s}^{i_1 \ldots i_r}$.

We shall determine the law of transformation of the components of a tensor under a change of base in V. In fact, let $\overline{e}_i$, $\overline{e'}^k$ be new dual bases in V, V^*, so that we have

$$\overline{e}_i = \sum_j \alpha_i^j e_j, \qquad \overline{e'}^k = \sum \beta_\ell^k e'. \tag{17}$$

Then

$$x = \sum_{(i)(k)} \xi_{k_1 \cdots k_s}^{-i_1 \cdots i_r} \overline{e}_{i_1} \cdots \overline{e}_{i_r} \overline{e'}^{k_1} \cdots \overline{e'}^{k_s}, \tag{18}$$

where $\overline{\xi}_{k_1 \cdots k_s}^{i_1 \cdots i_r}$ are the components of x relative to the new base $\overline{e}_i$ in V, substituting (17) into (18), we get

$$x = \sum_{(i)(j)(k)(\ell)} \overline{\xi}_{\ell_1 \ldots \ell_s}^{j_1 \ldots j_r} \alpha_{j_1}^{i_1} \cdots \alpha_{j_r}^{i_r} \beta_{k_1}^{\ell_1} \cdots \beta_{k_s}^{\ell_s} e_{i_1} \cdots e_{i_r} e'^{k_1} \cdots e'^{k_s}. \tag{19}$$

Comparison of (16) and (19) gives

$$\xi^{i_1\dots i_r}_{k_1\cdots k_s} = \sum_{(j)(\ell)} \overline{\xi}^{j_1\cdots j_r}_{\ell_1\dots\ell_s} \alpha^{i_1}_{j_1}\cdots\alpha^{i_r}_{j_r}\beta^{\ell_1}_{k_1}\cdots\beta^{\ell_s}_{k_s}. \tag{20}$$

The law of transformation (20) can be taken as a starting-point of the definition of a tensor, as was done in classical treatment.

Let u be an endomorphism of V. It induces an endomorphism u' of V^* defined by

$$\langle x, u'x'\rangle = \langle ux, x'\rangle, \qquad x \in V, x' \in V^*. \tag{21}$$

The tensor product

$$\underbrace{u\otimes\cdots\otimes u}_{r\ \text{fractors}}\otimes\underbrace{u'\otimes\cdots\otimes u'}_{s\ \text{fractors}}$$

which we write for simplicity by u^r_s, is then an endomorphism of V^r_s. By Theorem 4.1.13, if u is an automorphism of V, u^r_s is an automorphism of V^r_s; moreover, the map $u \longrightarrow u^r_s$ is a representation of the group of automorphisms of V into the group of automorphisms of V^r_s.

Definition 4.2.3. A linear subspace of V^r_s which is stable under the action of u^r_s for all u of the group of automorphisms of V is called a tensor space over V.

The space V^r_s itself is an example of a tensor space. We shall see more examples later on.

Definition 4.2.4. Given a Cartesian product

$$V\times\cdots\times V\times V^*\times\cdots\times V^*$$

with $r(\geq 1)$ factors V and $s(\geq 1)$ factors V^*, whose elements can be written $(x_1,\dots,x_r, x'_1,\dots,x'_s)$ where the x and the x' belong to V and V^* respectively. Let λ, μ be two indices such that $1 \leq \lambda \leq r$, $1 \leq \mu \leq s$. To the multilinear map

$$f_{\lambda\mu}: V\times\cdots\times V\times V^*\times\cdots\times V^* \longrightarrow V^{r-1}_{s-1} \tag{22}$$

defined by

$$f_{\lambda\mu}(x_1,\dots,x_r;x'_1,\dots,x'_s) = \langle x_\lambda, x'_\mu\rangle x_1\dots x_{\lambda-1}x_{\lambda+1}\dots x_r x'_1\dots x'_{\mu-1}x'_{\mu+1}\dots x'_s \tag{23}$$

corresponds the linear map

$$C_{\lambda\mu}: V^r_s \longrightarrow V^{r-1}_{s-1}. \tag{24}$$

This linear map is called a *contraction*.

Definition 4.2.5. Since $V^r_s\otimes V^{r'}_{s'}$ is canonically isomorphic with $V^{r+r'}_{s+s'}$, the bilinear map

$$V^r_s\times V^{r'}_{s'} \longrightarrow V^r_s\otimes V^{r'}_{s'}$$

gives rise to a bilinear map

$$V_s^r \times V_{s'}^{r'} \to V_{s+s'}^{r+r'}.$$

This operation is called *multiplication* of tensors. The product of a tensor of type (r,s) with a tensor of type (r',s') is a tensor of type $(r+r',s+s')$.

Definition 4.2.6. Let $T(V)$ be the (weak) direct sum $\sum_{r>0} V_0^r$. For

$$x = \sum_r x^r \in T(V), \qquad x^r \in V_0^r,$$

$$y = \sum_s y^s \in T(V), \qquad y^s \in V_0^s,$$

define their product

$$xy = \sum_{r,s} x^r \otimes y^s. \tag{25}$$

With this multiplication $T(V)$ becomes an algebra, called the *tensor algebra of* V.

We shall express the tensor operations, addition, contraction, and multiplication, in terms of the components relative to a base (14). Addition is of course component-wise. If $x \in V$ is given by (16), the components of $C_{\lambda\mu}(x)$ are

$$\sum_k \xi^{i_1 \cdots i_{\lambda-1} k i_{\lambda+1} \cdots i_r}_{j_1 \cdots j_{\mu-1} k j_{\mu+1} \cdots j_s} \tag{26}$$

where the summation is with respect to the λ-th upper index and the μ-th lower index. If y has the components

$$\eta^{k_1 \ldots k_{r'}}_{\ell_1 \ldots \ell_{s'}},$$

the components of $x \otimes y$ are

$$\xi^{i_1 \cdots i_r k_1 \cdots k_{r'}}_{j_1 \cdots j_s \ell_1 \ldots \ell_{s'}} = \xi^{i_1 \ldots i_r}_{j_1 \ldots j_s} \eta^{k_1 \ldots k_{r'}}_{\ell_1 \ldots \ell_{s'}}. \tag{27}$$

Perhaps the main advantage of the component notation is that in practice there are different tensor spaces of order (r,s) depending on the order of the V's and the $V^{*'}$s, and we may not be allowed to identify them. Thus, $V \otimes V^* \otimes V$ and $V \otimes V \otimes V^*$ are different tensor spaces and in general should not be identified. In the component notation their elements can be distinguished by the notations $\xi^{i \cdot k}_{j}$ and $\xi^{ij \cdot}_{k}$ respectively.

4.3 Symmetry and skew-symmetry; Exterior algebra

Let V be a vector space and let us consider the space V_0^r of its contravariant tensors of order r. For simplicity we shall write V^r for V_0^r. Denote by

$$h : \underbrace{V \times \cdots \times V}_{r \text{ times}} \to V^r$$

the natural projection. Let σ be a permutation on r letters. To τ we can define the r-linear map

$$f_\sigma : \underbrace{V \times \cdots \times V}_{r \text{ times}} \to V^r$$

by

$$f_\sigma(x_1, \ldots, x_r) = x_{\sigma(1)} \otimes x_{\sigma(2)} \otimes \cdots \otimes x_{\sigma(r)}.$$

There exists then a linear map $V^r \to V^r$, which we shall also denote by σ, such that $f_\sigma = \sigma h$. This linear map σ satisfies the relation

$$\sigma(x_1 \otimes \cdots \otimes x_r) = x_{\sigma(1)} \otimes \cdots \otimes x_{\sigma(r)}. \tag{28}$$

Definition 4.3.1. A tensor $x \in V^r$ is called *symmetric* if $\sigma x = x$ for every permutation. It is called *anti-symmetric* (or skew-symmetric) or alternate if $\sigma x = (\operatorname{sgn}\sigma)x$ for every permutation σ, where $\operatorname{sgn}\sigma$ is $+1$ or -1 according as the permutation σ is even or odd.

Definition 4.3.2. A linear map $f : V^r \to Z$, where Z is a vector space, is said to be *symmetric* if $f\sigma = f$ for all permutations σ, that is, if $f\sigma(y) = f(y)$ for all $y \in V^r$ and all σ. It is said to be alternate if $f\sigma = (\operatorname{sgn}\sigma)f$ for all σ.

For $z \in V^r$ we introduce the operates

$$\begin{aligned} S_r(z) &= \sum \sigma(z), \\ A_r(z) &= \sum_\sigma (\operatorname{sgn}\,\sigma)\sigma(z), \end{aligned} \tag{29}$$

the summations being over all permutations σ of $1, \ldots, r$. They define linear maps of V into itself, to be called respectively *symmetrization* and *alternation*. The names are justified by the following theorem:

Theorem 4.3.3. *For any $z \in V^r$, $S_r(z)$ is symmetric and $A_r(z)$ is alternate. Moreover,*

$$\begin{aligned} S_r(z) &= r|z, \quad \text{if } z \text{ is symmetric}, \\ A_r(z) &= r|z, \quad \text{if } z \text{ is alternate}. \end{aligned} \tag{30}$$

We shall prove the statement for $A_r(z)$, the proof of the other part being similar. For any permutation τ on $1, \ldots, r$, we have

$$\tau A_r(z) = \sum_\sigma (\operatorname{sgn}\,\sigma)\tau\sigma(z) = \operatorname{sgn}\,\tau \sum_\sigma (\operatorname{sgn}\,\tau\sigma)\tau\sigma(z) = (\operatorname{sgn}\,\tau)A_r(z).$$

In case z is alternate, we have

$$A_r(z) = \sum_\sigma (\operatorname{sgn}\,\sigma)^2 z = r|z.$$

Hence the theorem is proved.

Definition 4.3.4. The quotient space $\wedge^r(V) = V^r/N^T$, where N^r is the kernel of the linear map $A'_r V^r \to V^r$, is called the space of exterior vectors over V. Its elements are called *exterior r-vectors* or simply *r-vectors*. This definition has a meaning only when $r \geq 2$. By convention we define $\wedge^1(V) = V$, $\wedge^0(V) = F$.

By Theorem 4.3.3 the image of A_r is the space of alternate (contravariant) r-tensors. Hence A_r induces on isomorphism of the space $\wedge^r(V)$ of exterior r-vectors onto the space of alternate r-tensors.

If $\psi : V^r \to \wedge^r(V)$ is the natural projection, sending an element of V^r to its coset mod N^r we shall use the notation

$$\psi(x_1 \otimes \cdots \otimes x_r) = x_1 \wedge x_2 \wedge \cdots \wedge x_r, \tag{31}$$

for any $x_1, \ldots, x_r \in V$. An r-vector of the form (31) is called *decomposable*.

Theorem 4.3.5. *A linear map $f : V^r \to Z$, where Z is a vector space, is alternate if and only if $f(N^r) = 0$.*

Let f be alternate and let $y \in N^r$, so that $A_r(y) = 0$. The hypothesis that f is alternate gives

$$f(y) = (\operatorname{sgn} \sigma) f\sigma(y).$$

It follows that

$$\begin{aligned} r|f(y) &= \sum_\sigma (\operatorname{sgn} \sigma) f\sigma(y) = f\left\{ \sum_\sigma (\operatorname{sgn} \sigma)\sigma(y) \right\} \\ &= fA_r(y) = 0, \end{aligned}$$

where the summations are over all permutations σ of $1, \ldots, r$.

Conversely, suppose $f(N^r) = 0$. For any $y \in V^r$ and any permutation σ, we have

$$A_r\{y - (\operatorname{sgn} \sigma)\sigma(y)\} = \sum_\tau (\operatorname{sgn} \tau)\tau(y) - (\operatorname{sgn} \sigma) \sum_\tau (\operatorname{sgn} \tau) t\sigma(y)$$

which implies that

$$y - (\operatorname{sgn} \sigma)\sigma(y) \in N^r.$$

Hence

$$f\{y - (\operatorname{sgn} \sigma)\sigma(y)\} = 0$$

or

$$f(y) = (\operatorname{sgn} \sigma) f\sigma(y).$$

Thus the theorem is proved.

Corollary 4.3.6. *The space of all alternate linear maps $f : V^r \to Z$ is isomorphic with the space of all linear maps $f : \wedge^r(V) \to Z$, the correspondence being established by the relation $f = g\psi$.*

By Theorem 4.3.5, ψ is alternate, whence $g\psi$ is alternate. The map which sends g into $g\psi$ is clearly a linear map of the space of linear maps g into the space of alternate linear maps f. From Theorem 4.3.5 it follows that a given f induces a linear map g such that f can be factored in the form $f = g\psi$. Moreover, since ψ is onto, $f = 0$ if and only if $g = 0$. This proves the corollary.

Let V, W be vector spaces, and let $u : V \longrightarrow W$ be a linear map. Denote by $u^r : V^r \longrightarrow W^r$ the tensor power of u (def. 4.1.12) and by $\eta : W^r \longrightarrow \wedge^r(W)$ the natural projection. Then $\eta u^r : V^r \longrightarrow \wedge^r(W)$ is a linear map, and we shall prove that it is alternate. In fact, for any permutation σ of $1, \ldots, r$, we have

$$u^r \sigma = \sigma u^r.$$

Since η is alternate, it follows that

$$\eta u^r \sigma = \eta \sigma u^r = (\text{sgn } \sigma) \eta u^r,$$

which proves that ηu^r is alternate. By the above Corollary there exists a linear map $\wedge^r u : \wedge^r(V) \longrightarrow \wedge^r(W)$, such that $(\wedge^r u) \circ \psi = \eta \circ u^r$. We call $\wedge^r u$ the rth *exterior power* of u. For this linear map we have the relation.

$$(\wedge^r u)(x_1 \wedge \cdots \wedge x_r) = u(x_1) \wedge \cdots \wedge u(x_r). \tag{32}$$

We shall now define an important operation, to be called *exterior multiplication*, which pairs an element of $\wedge^r(V)$ and an element of $\wedge^s(V)$ into an element of $\wedge^{r+s}(V)$. When r or s is O, $\wedge^0(V)$ is by definition F and we define the operation to be scalar multiplication. Suppose $r, s \geq 1$. The tensor product $V^r \otimes V^s$ can be identified with V^{r+s}, so that we have the sequence of maps

$$V^r \times V^s \xrightarrow{h} V^r \otimes V^s \xrightarrow{i} V^{r+s} \xrightarrow{\psi} \wedge^{r+s}(V).$$

Of these h is bilinear, i is an isomorphism, identifying the two vector spaces and ψ is the natural projection. The product $u = \psi i h$ is therefore bilinear. Moreover, $u(x,y)$, $x \in V^r$, $y \in V^s$, defines an alternate linear map

$$u_y : V^r \longrightarrow \wedge^{r+s}(V)$$

by

$$u_y(x) = u(x,y)$$

for any fixed $y \in V^s$, and an alternate linear map

$$u_x : V^s \longrightarrow \wedge^{r+s}(V)$$

by

$$u_x(y) = u(x,y)$$

for any fixed $x \in V^r$. It follows from Theorem 4.3.5 that

$$u(x,y)=0, \qquad x\in N^r,$$
$$u(x,y)=0, \qquad y\in N^s.$$

The map u therefore induces a bilinear map

$$\tilde{u}:\wedge^r(V)\times\wedge^s(V)\longrightarrow\wedge^{r+s}(V) \tag{33}$$

defined by

$$u(\xi,\eta)=u(x,y), \quad \xi\in\wedge^r(V), \quad \eta\in\wedge^s(V),$$

where x, y are two representatives in the classes ξ, η defined mod N^r, N^s respectively.

Definition 4.3.7. The element $\tilde{u}(\xi,\eta)\in\wedge^{r+s}(V)$, for $\xi\in\wedge^r(V)$, $\eta\in\Delta^s(V)$, is called the *exterior product* of ξ, η. It will be denoted by $\xi\wedge\eta$.

Theorem 4.3.8. *Exterior multiplication has the following properties:*

$$(e_1\wedge\cdots\wedge e_r)(f_1\wedge\cdots\wedge f_s)=e_1\wedge\cdots\wedge e_r\wedge f_1\wedge\cdots\wedge f_s, \tag{34}$$
$$(\alpha_1x_1+\alpha_2x_2)\wedge y=\alpha_1(x_1\wedge y)+\alpha_2(x_2\wedge y), \tag{35}$$
$$x\wedge(\alpha_1y_1+\alpha_2y_2)=\alpha_1(x\wedge y_1)+\alpha_2(x\wedge y_2),$$
$$(x\wedge y)\wedge z=x\wedge(y\wedge z), \tag{36}$$
$$x\wedge y=(-1)^{rs}y\wedge x, \tag{37}$$

where α_1, α_2 are scalars, $e_1,\dots,e_r$, $f_1,\dots,f_s$ are vectors and s, x_1, x_2, y, y_1, y_2, z are exterior multivectors with $r=$ order of x, $s=$ order of y.

Formula (34) follows from definition, while (35) follows from the bilinearity of the map u. Formula (36) is true for decomposable multivectors, in view of (34). Its validity for general multivectors then follows by linearity. This "associativity" of exterior multiplication allows us to write the common expression in (36) as $x\wedge y\wedge z$.

To prove (37) we notice that by definition

$$e_1\wedge\cdots\wedge e_r=(\operatorname{sgn}\sigma)e_{\sigma(1)}\cdots e_{\sigma(r)}. \tag{38}$$

It follows immediately that (37) is true when x and y are decomposable. By linearity the relation is extended to general multivectors.

Corollary 4.3.9. *For $r>n=\dim V$, $\wedge^r(V)=0$.*

This is immediate consequence of (37).

Definition 4.3.10. We consider the direct sum

$$\wedge(V)=\sum_{0\le r\le n}\wedge^r(V). \tag{39}$$

For

$$x = \sum_{0 \le r \le n} x^r, y = \sum_{0 \le s \le n} y^s, x^r \in \wedge^r(V), y^s \in \wedge^s(V) \tag{40}$$

define their product by

$$x \wedge y = \sum_{r,s} x^r \wedge y^s. \tag{41}$$

Then $\wedge(V)$ becomes an algebra, called the exterior or Grassmann algebra of the vector space V.

Theorem 4.3.11. *The exterior algebra $\wedge(V)$ of a vector space V of dimension n is of dimension 2^n. If $e_1, \ldots, e_n$ form a base of V, a base of $\wedge(V)$ is given by 1 and the elements*

$$e_{i_1} \wedge \cdots \wedge e_{i_r}, \quad i_1 < i_2 < \ldots < i_r, \quad i_1, \ldots, i_r = 1, \ldots, n, \quad r = 1, \ldots, n. \tag{42}$$

We first observe that by the definition of alternation,

$$A_n(e_1 \otimes \cdots \otimes e_n) \neq 0,$$

which gives

$$\psi(e_1 \otimes \cdots \otimes e_n) = e_1 \wedge \cdots \wedge e_n \neq 0. \tag{43}$$

Suppose there be a linear relation of the form

$$\sum_{i_1 < \cdots < i_r} \alpha_{i_1 \cdots i_r} e_{i_1} \wedge \cdots \wedge e_{i_r} = 0,$$

where the summation is over all combinations $i_1, \ldots, i_r$ of $1, \ldots, n$. For a fixed set of indices $i_1, \ldots, i_r$, let $j_1, \ldots, j_{n-r}$ be a complementary set and multiply the equation by $e_{j_1} \wedge \cdots \wedge e_{j_{n-r}}$. With the exception of the term in $e_{i_1} \wedge \cdots \wedge e_{i_r}$ all other terms in the product will have repeated factors and will be zero by means of (37). Hence we get

$$\alpha_{i_1 \ldots i_r} e_{i_1} \wedge \cdots \wedge e_{i_r} e_{j_1} \wedge \cdots \wedge e_{j_{n-r}} = 0.$$

Using (43) it follows that

$$\alpha_{i_1 \ldots i_r} = 0,$$

which proves that the elements in (42) are linearly independent. On the other hand, it is clear that an element of $\wedge(V)$ is a linear combination of the elements of (42) Hence the theorem is proved.

In terms of the base (42) an element of $\wedge^r(V)$ can be written in the form

$$x = \sum \alpha^{i_1 \ldots i_r} e_{i_1} \wedge \cdots \wedge e_{i_r}. \tag{44}$$

Instead of taking the sum over all combinations of $i_1, \ldots, i_r$, we find it more convenient to take the sum for $i_1, \ldots, i_r$ running independently from 1 to n and assume that the coefficients $\alpha_{i_1 \cdots i_r}$ are anti-symmetric in every pair of indices. These coefficients are thus uniquely determined and will be called the components of the r-vector relative to the base e_i of V.

We proceed to give some geometrical applications of the exterior algebra, as follows:

Theorem 4.3.12. *A necessary and sufficient condition that the vectors $x_1,\dots,x_r$ be linearly dependent is their exterior product be zero:*

$$x_1 \wedge \cdots \wedge x_r = 0. \tag{45}$$

If the vectors in question are linearly dependant, one of them, say x_r, is a linear combination of the others:

$$x_r = \sum_{k=1}^{r-1} \alpha_k x_k.$$

Hence we have

$$x_1 \wedge \cdots \wedge x_r = x_1 \wedge \cdots \wedge x_{r-1} \wedge \left(\sum_{k=1}^{r-1} \alpha_k x_k \right) = \sum_{k=1}^{r-1} \alpha_k (x_1 \wedge \cdots \wedge x_{r-1} \wedge x_k).$$

Each summand of the last sum has two repeated factors and is therefore zero.

Suppose next that (45) holds and that linearly independent. Then $x_{r+1},\dots,x_n$ can be chosen so that $x_1,\dots,x_n$ from a base in V. By Theorem 4.3.11, we have

$$x_1 \wedge \cdots \wedge x_n \neq 0,$$

which contradicts (45).

Theorem 4.3.13. *To a linear subspace W of dimension r of a vector space V (of dimension n) corresponds a decomposable r-vector $w \neq 0$, determined up to a scalar factor, such that the vectors $x \in W$ are characterized by the property $x \wedge w = 0$.*

Let W_1, W_2 be two linear subspaces, of dimensions r_1 and r_2 respectively, and let w_1, w_2 be the corresponding r_1-vector and r_2-vector. In order that $W_1 \subset W_2$ it is necessary and sufficient that a $(r_2 - r_1)$-vector v exists such that $w_2 = v \wedge w_1$. In order that W_1, W_2 have an empty intersection it is necessary and sufficient that $w_1 \wedge w_2 \neq 0$. Moreover, if this condition is satisfied, $w_1 \wedge w_2$ is the $(r_1 + r_2)$-vector corresponding to the ambiant space of W_1, W_2.

The r-vector w is defined by

$$w = y_1 \wedge \cdots \wedge y_r, \tag{46}$$

where $y_1,\dots,y_r$ form a base in W. If we take a different base $y'_1,\dots,y'_r$ in W and put

$$w' = y'_1 \wedge \cdots \wedge y'_r,$$

we have

$$y'_k = \sum \alpha_{k\ell} y_\ell, k, \quad \ell = 1,\dots,r,$$

and

$$w' = |\alpha_{k\ell}| w,$$

where the first factor is the determinant. Hence w is determined up to a scalar factor. A vector $x \in W$ if and only if it is linearly dependent on the vectors of a base, that is, if and only if $x \wedge w = 0$.

We leave the proof of the second part of the theorem to the reader.

Before concluding this section, we wish to remark that, using the symmetrization, an algebra can be defined in an analogous way. In fact, denote by M^r the kernel of the linear map $S_r : V^r \to V^r$, and put $P^r(V) = V^r/M^r$. Then a bilinear map

$$P^r(V) \times P^s(V) \to P^{r+s}(V) \tag{47}$$

can be defined, which is an analogue of the exterior product. Using this product, a multiplication can be defined in the direct sum

$$P(V) = \sum_{r \geq 0} P^r(V). \tag{48}$$

This multiplication makes $P(V)$ into an algebra, called the *symmetric algebra* of V. Unlike the exterior algebra the symmetric algebra is infinite-dimensional. It is playing an increasingly important role in geometry.

4.4 Duality in exterior algebra

By Corollary 4.1.8 to Theorem 4.1.7 we established a canonical isomorphism between $V^* \otimes W^*$ and $(V \otimes W)^*$. We shall identify these two spaces by means of this isomorphism. Then we can write

$$\begin{gathered} \langle x \otimes y, x' \otimes y' \rangle = \langle x, x' \rangle \langle y, y' \rangle, \\ x \in V, \quad y \in W, \quad x' \in V^*, \quad y' \in W^*. \end{gathered} \tag{49}$$

This result can be extended to any finite number of factors. In particular, we can identify $(V^r)^*$ with

$$V_r = \underbrace{V^* \quad \cdots \quad V^*}_{r \text{ times}}. \tag{50}$$

The fundamental bilinear relation, between V^r and (50) is given by

$$\langle x_1 \otimes \cdots \otimes x_r, x'_1 \otimes \cdots \otimes x'_r \rangle = \langle x_1, x'_1 \rangle \cdots \langle x_r, x'_r \rangle. \tag{51}$$

We proceed to make a similar study for the exterior algebra, identifying $(\wedge^r(V))^*$ and $\wedge^r(V^*)$. Denote $\mathscr{A}(V^r;F)$ the vector space of alternate linear maps $V^r \to F$, and by $\mathscr{A}(V_r)$ the vector space of alternate covariant r-tensors. By the Corollary (4.3.6), we have the canonical isomorphism

$$(\wedge^r(V))^* = \mathscr{L}(\wedge^r(V);F) \cong \mathscr{A}(V^r;F). \tag{52}$$

On the other hand, by the remark following Theorem 4.3.3, applied to the vector space V^*, we have the isomorphism.

$$\wedge^r(V^*) \cong \mathscr{A}(V_r). \tag{53}$$

We shall prove that the right-hand sides of (52) and (53) are canonically isomorphic. In fact, let $\vartheta : V^r \to F$ be an alternate linear map. We define an element $z'(\vartheta) \in V_r$ by the equation

$$\langle x_1 \otimes \cdots \otimes x_r, z'(\vartheta)\rangle = \vartheta(x_1 \otimes \cdots \otimes x_r),\ x_1, \ldots, x_r \in V. \tag{54}$$

For any permutation σ on $1, \ldots, r$ we have

$$\langle x_1 \otimes \cdots \otimes x_r, \sigma z'(\vartheta)\rangle = \langle x_{\sigma^{-1}(1)} \otimes \cdots \otimes x_{\sigma-1(r)}, z'(\vartheta)\rangle$$
$$\vartheta(x_{\sigma^{-1}(1)} \otimes \cdots \otimes x_{\sigma^{-1}(r)}) = \operatorname{sgn}(\sigma^{-1})\vartheta(x_1 \otimes \cdots \otimes x_r) = (\operatorname{sgn}\sigma)\vartheta(x_1 \otimes \cdots \otimes x_r).$$

It follows that

$$\sigma z'(\vartheta) = (\operatorname{sgn}\sigma) z'(\vartheta),$$

that is, $z'(\vartheta)$ is alternate. The map defined by sending ϑ to $z'(\vartheta)$ is a linear map $z' : \mathscr{A}(V^r; F) \to \mathscr{A}(V_r)$, and we see easily that it is a one-one isomorphism. It follows that any two vector spaces in the formulas (52), (53) are canonically isomorphic.

To express these isomorphisms more explicitly, Let us denote by k and $\bar{k}$ respectively the isomorphisms in (52), (53), being each time the linear map of the vector space in the left-hand side to that of the right-hand side. Then we have

$$\langle x_1 \wedge \cdots \wedge w_r, k^{-1}(\vartheta)\rangle = \vartheta(x_1 \otimes \cdots \otimes x_r). \tag{55}$$

On the other hand, the isomorphism in (53) gives, $x'_1, \ldots, x'_r$ being vectors of V^*,

$$\begin{aligned}\langle x_1 \otimes \cdots \otimes x_r, \bar{k}(x'_1 \wedge \cdots \wedge x'_r)\rangle &= \langle x_1 \otimes \cdots \otimes x_r, A_r(x'_1 \otimes \cdots \otimes x'_r)\rangle \\ &= \left\langle x_1 \otimes \cdots \otimes x_r, \sum_\sigma (\operatorname{sgn}\sigma) x'_{\sigma(1)} \otimes \cdots \otimes x'_{\sigma(r)}\right\rangle \\ &= \sum_\sigma (\operatorname{sgn}\sigma)\langle x_1, x'_{\sigma(1)}\rangle \cdots \langle x_r, x'_{\sigma(r)}\rangle \\ &= |\langle x_i, x'_k\rangle|.\end{aligned} \tag{56}$$

We now put

$$\bar{k}(x'_1 \wedge \cdots \wedge x'_r) = z'(\vartheta). \tag{57}$$

By (54), (55), (56), (57), we get

$$\langle x_1 \wedge \cdots \wedge x_r, k^{-1} z'^{-1} \bar{k}(x'_1 \wedge \cdots \wedge x'_r)\rangle = |\langle x_i, x'_k\rangle|.$$

When we say that we identify the vector spaces by the canonical isomorphisms, we mean exactly that the isomorphism $k^{-1}z'^{-1}\overline{k}$ can be dropped. As a result we get

$$\langle x_1 \wedge \cdots \wedge x_r, x'_1 \wedge \cdots \wedge x'_r \rangle = |\langle x_i, x'_k \rangle|. \tag{58}$$

We summarize our results in the following theorem:

Theorem 4.4.1. *The spaces $(\wedge^r(V))^*$ and $\wedge^r(V^*)$ are isomorphic. They are identified by this isomorphism, the pairing of $\wedge^r(V)$ and $\wedge^r(V^*)$ into F is given by* (58).

Definition 4.4.2. The vector space $(\wedge^r(V))^* = \wedge^r(V^*)$ is called the space of exterior over V. An element of this space is called an r-form, $\wedge(V^*)$ is called the algebra of exterior forms on V.

Theorem 4.4.3. *Let $u : V \longrightarrow W$ be a linear map, and $u' : W^* \longrightarrow V^*$ its dual map. Then the linear maps $\wedge^r u : \wedge^r(V) \longrightarrow \wedge^r(W)$ and $\wedge^r u' : \wedge^r(W^*) \longrightarrow \wedge^r(V^*)$ are dual to each other.*

We leave the proof of this theorem to the reader.

Definition 4.4.4. Let $x \in \wedge^r(V)$, $x' \in \wedge^s(V^*)$. For $r \leq s$ the left interior product $x \rfloor x' \in \wedge^{s-r}(V^*)$ is defined by

$$\langle z, x \rfloor x' \rangle = \langle z \wedge x, x' \rangle z \in \wedge^{s-r}(V). \tag{59}$$

For $s \leq r$ the right interior product $x \lfloor x' \in \wedge^{r-s} V$ is defined by

$$\langle z, x \lfloor x', z' \rangle = \langle x, x' \wedge z' \rangle, \qquad z' \in \wedge^{r-s}(V^*). \tag{60}$$

We define the left interior product to be 0 when $r > s$ and the right interior product to be 0 when $s > r$.

In particular, when $r = s$, we have

$$x \rfloor x' = x \lfloor x' = \langle x, x' \rangle. \tag{61}$$

The use of the interior product is to establish relations between the algebras $\wedge(V)$ and $\wedge(V^*)$. As an example we shall prove the following theorem:

Theorem 4.4.5. *Let e, e' be dual bases of the one-dimensional vector spaces $\wedge^n(V)$ and $\wedge^n(V^*)$, n being the dimension of V. The linear map: $\varphi : \wedge^r(V) \longrightarrow \wedge^{n-r}(V^*)$ defined by*

$$\varphi(x) = x \rfloor e', \quad x \in \wedge^r(V) \tag{62}$$

is a one-one isomorphism and carries decomposable elements into decomposable elements. If $x \neq 0$ is decomposable and defines the linear subspace W of dimension r, then $\varphi(x) \neq 0$ defines the linear subspace of orthogonal to W.

We choose dual bases $e_1,\dots,e_n$ and $e'_1,\dots,e'_n$ of V and V^* respectively, such that

$$e = e_1 \wedge \cdots \wedge e_n, \; e' = e'_1 \wedge \cdots \wedge e'_n. \tag{63}$$

Then by definition

$$(e_{i_1} \wedge \cdots \wedge e_{i_r}) \rfloor (e'_1 \wedge \cdots \wedge e'_n) = \varepsilon(i_1,\dots,i_r) e'_{j_1} \wedge \cdots \wedge e'_{j_{n-r}}. \tag{64}$$

where $j_1,\dots,j_{n-r}$ form a set of indices complementary to $i,\dots,i$ and $\varepsilon(i_1 \dots i_r)$ is $+1$ or -1 according as the permutation which brings $j_1,\dots,j_{n+1}, i_1,\dots,i_r$ to the order $1,\dots,n$ is even or odd. From this it follows immediately that the map φ is a one-one isomorphism.

When the given element is decomposable, we can, by a suitable choice of the base in V, assume it be $e_1 \wedge \cdots \wedge e_r$. Then $\varphi(e_1 \wedge \cdots \wedge e_r) = \pm e'_{r+1} \wedge \cdots \wedge e'_n$ and the second statement is clear.

Remark 4.4.6. *The definition of the isomorphism φ depends on the choice of the base e in $\wedge^n(V)$. Multiplication of e by a non-zero scalar factor changes φ accordingly.*

Corollary 4.4.7. *Every $(n-1)$-vector of a vector space of n dimension is decomposable.*

4.5 Inner product

In this section we assume that F is either the real field or the complex field. An inner product is a function (x,y) defined in $V \times V$, with values in F, satisfying the conditions:

1) $(x,y) = (\overline{y},\overline{x})$, where the bars over a scalar denotes its conjugate complex number;
2) $(\alpha_1 x_1 + \alpha_2 x_2, y) = \alpha_1(x_1,y) + \alpha_2(x_2,y)$, $\alpha, \alpha_2 \in F$, $x_1, x_2, y \in V$;
3) (x,x), being real as a consequence of 1), is ≥ 0, and is $= 0$ if and only if $x = 0$.

It is well-known that the correspondence between $y \in V$ and $y' \in V^*$ established by the condition

$$\langle x, y' \rangle = (x,y), \; x \in V \tag{65}$$

defines a one-one conjugate isomorphism between V and V^*.

Theorem 4.5.1. *An inner product in V induces inner products in V^r, $\wedge^r(V)$, and $P^r(V)$.*

We define in V^r,

$$(x_1 \otimes \cdots \otimes x_r, y_1 \otimes \cdots \otimes y_r) = (x_1,y_1)\cdots(x_r,y_r). \tag{66}$$

It is obvious that this function satisfies the conditions of an inner product.

Restricting this function to the subspace of alternate elements $\mathscr{A}(V^r)$ and the subspace of symmetric elements $S(V^r)$ of V^r, we get inner products in these spaces. Since the latter are isomorphic to $\wedge^r(V)$ and $P^r(V)$ respectively, these inner products give rise to inner products of $\wedge^r(V)$ and $P^r(V)$.

Theorem 4.5.2. *In a real vector space of dimension n with an inner product there exists a base $e \in \wedge^n(V)$, determined up to signs, such that $(e,e) = 1$.*

For any base element $\tilde{e}$ of $\wedge^n(V)$ we have

$$(\tilde{e},\tilde{e}) = \alpha > 0.$$

We replace $\tilde{e}$ by $e = \pm\frac{e}{\sqrt{\alpha}}$. Then e satisfies the conditions of the theorem, and is determined up to a sign.

Remark 4.5.3. *A choice between e and $-e$ is called an orientation of V.*

Applying Theorem 4.4.5 we have therefore in an orient real vector space a well-defined isomorphism

$$\wedge^r(V) \cong \wedge^{n-r}(V^*). \tag{67}$$

On the other hand., the inner products establish the isomorphisms

$$\wedge^r(V) \cong \wedge^r(V^*),$$
$$\wedge^{n-r}(V) \cong \wedge^{n-r}(V^*).$$

The isomorphism between $\bigwedge^r(V^*)$ and $\bigwedge^{n-r}(V^*)$ obtained by combining these isomorphisms is the star operation of Hodge, which plays an important role in the theory of harmonic integrals.

5

Differentiable Manifolds

5.1 Definition of a differentiable manifold

A differentiable manifold is, roughly speaking, a space in which differentiation has a sense. This will be achieved by selecting a family of real-valued functions in the space and ascribing to them properties to be possessed by differentiable functions.

Definition 5.1.1. A differentiable manifold is a separable (i.e., with a countable base), Hausdorff space of M with a family $F_k (1 \leq k \leq +\infty)$ of real-valued functions, defined on open subsets M, such that the following conditions are satisfied:

1) Every function f, whose domain of definition U in M is the union of a family of open sets U_α, belongs to F_k, if its restriction $f|U_\alpha$ to each U_α belongs to F_k. Conversely the restriction of a function of F_k to an open subset V of its domain of definition U belongs to F_k.
2) To every point $p \in M$ there is a homeomorphism u of an open subset U of M containing p to an open subset $u(U)$ of an n-dimensional Euclidean space E^n, such that, for every open subset $V \subset U$ with $p \in V$, the family of all functions $\{f\}$ of F_k where domain of definition is V is identical with the family $\{g \circ u\}$, where g runs over all functions of class k in $u(V)$. (A function in an open subset V' of E^n is said to be of class k, if it has at every point of V' all partial derivatives of order $\leq k$, those of order k being moreover continuous).

The functions of the family F_k are called *functions of class k* on M. They are said to define a *differentiable structure* of class k on M. The integer n is called the *dimension* of the differentiable manifold, being also the dimension of M as a topological space. The space M is called the *underlying topological space*. The differentiable manifold will be denoted by $\{M, F_k\}$, and also M when there is no danger of confusion.

When, in Condition (2) of the above definition, we allow g to run over only the analytic functions in $u(V)$, that is, functions which at every point of their domains of definition can be expanded into convergent power series, the manifold, to be denoted by $\{M,F_w\}$, is called *analytic*.

Let $\{M,F_k\}$ be a given differentiable manifold of class k. It follows from Condition (2) that to every point $p\in M$ there is a neighbourhood U of p and a homeomorphism $u:U\to u(U)\subset E^n$ such that any function of class k defined in an open subset V, $p\in V\subset U$, is of the form $g\circ U$, where g is of class k in $u(V)$. Let u defined by

$$u(q)=(u^1(q),\dots,u^n(q)),\qquad q\in V. \tag{1}$$

Then we can write

$$g\circ u(q)=g(u^1(q),\dots,u^n(q)),\qquad q\in V. \tag{2}$$

The functions $u^1(q),\dots,u^n(q)$, which obviously belong to F_k, are called the *local coordinates* in U or the *local coordinates at p*.

Suppose that, instead of U, there exists a different neighbourhood U' of p, with the homeomorphism $u':U'\to u'(U')$, having the same properties. By restricting to a smaller neighbourhood when necessary, we assume $U=U'$. Then $u'\circ u^{-1}$ is a homeomorphism of $u(U)$ onto $u'(U)$ and can be defined by the equations

$$u'^i=u'^i(u^1,\dots,u^n),\qquad i=1,\dots,n, \tag{3}$$

where the functions in the right hand side are of class k, since $u'^1(q),\dots,u'^n(q)$ are functions of class k in U. A function of class k in V can now be written in one of the two forms:

$$f=g\circ u=g'\circ u', \tag{4}$$

and we have

$$g=g'\circ u^1\circ u^{-1} \tag{5}$$

or

$$g(u^1,\dots,u^n)=g'(u'^1(u^1,\dots,u^n),\dots,u'^n(u^1,\dots,u^n)). \tag{6}$$

This formula gives the relation between two "representations" g, g' of a function of class k in V in terms of two systems of local coordinates.

From the family of functions F_k, we can select a sub-family H_k having the following property:

(P) To every point $p\in M$ there is a neighbourhood U containing p in which there are functions $u^1,\dots,u^n$ of H_k, with the following properties:

1) The mapping $u:U\to u(U)$ of U onto an open subset $u(U)$ of E^n defined by

$$u(q)=(u^1(q),\dots,u^n(q)),\qquad q\in U$$

is a homeomorphism.

2) If $f\in H_k$ is defined at p, there is an open subset V of U containing p such that f/V is identical with $g\circ u$, g being a function of class k in $u(V)$.

The significance of such a sub-family lies in the fact that it is possible to determine the family F_k from the knowledge of the family H_k, as given by the theorem:

Theorem 5.1.2. *Let M be a separable Hausdorff space with a family H_k of real-valued functions, defined on open subsets of M, such that (P) holds. There exists one, and only one, minimal family of functions F_k, defined on open subsets of M, which contains H_k as a sub-family, such that $\{M, F_k\}$ is a differentiable manifold.*

Consider a real-valued function f defined on an opensubset W of M. Let $p \in W$, with the neighbourhood U having properties postulated by (P). Suppose that there is a neighbourhood V of p contained in $U \cap W$ such that f/V is identical with $g \circ u$, where g is a function of class k in $u(V)$. If f has this property at every point $p \in W$, it clearly belongs to any family F_k of functions which contains H_k and defines a differentiable structure of class k on M.

On the other hand, let F_k be the family of all these functions f. The conditions 1), 2) of Definition 5.1.1 are easily verified, so that $\{M, F_k\}$ is a differentiable manifold of class k.

Corollary 5.1.3. *A differentiable structure of class k on M defines a (minimal) differentiable structure of class $\ell < k$, by the condition that the functions of class k are also functions of class ℓ.*

Instead of the family of functions F_k it is sometimes convenient to use a family H'_k which also determines the differentiable structure completely and which has the following property:

(P') There exists a countable open covering $\{U_\alpha\}$ of M such that to each α there are n functions $u^1_\alpha, \ldots, u^n_\alpha$ of H'_k, with U_α as their domain of definition, having the property that the mapping $u_\alpha : U_\alpha \to E^n$ defined by

$$u_\alpha(q) = (u^1_\alpha(q), \ldots, u^n_\alpha(q)), \qquad q \in U_\alpha,$$

is a homeomorphism of U_α onto an open subset $u_\alpha(U_\alpha)$ of E^n. Moreover, if $f \in H'_k$ is defined at p and $p \in U_\alpha$, there is on open subset V of U_α containing p such that f/V is identical with $g \circ u_\alpha$, where g is a function of class k in $u_\alpha(V)$.

On account of the above results it is sufficient to define a differentiable structure by giving the family of functions H_k or H'_k we illustrate this by the examples:

Example 5.1.4. The family of functions $H'_k = \{x^1, \ldots, x^p\}$ in E^n where $x_1, \ldots, x^n$ are the coordinates, defines a differentiable structure on E^n.

Example 5.1.5. M is the unit $(n-1)$-sphere in E^n defined by the equation

$$(x^1)^2 + \cdots + (x^n)^2 = 1.$$

The family of functions $H'_k = \{x^1, \ldots, x^n\}$ defines a differentiable structure on M.

Example 5.1.6. M is the Euclidean line E^1 with the coordinate x^1 and H'_k consists of the function $(x^1)^3$. This structure on E' is different from the one in Example 5.4.2, where there are more functions of class k.

Theorem 5.1.7. The underlying space M of a differentiable manifold $\{M, F_k\}$ has a countable open covering $\{U_\alpha\}$ with the following properties:

1) *To each* α *there is a homeomorphism* $u_\alpha : U_\alpha \to E^n$.
2) *If* $p \in U_\alpha \cap U_\beta$, *there a neighbourhood* V *of* p *such that each coordinate of the point* $u_\beta u_\alpha^{-1}(q')$, $q' \in u_\alpha(V)$, *is a function of class* k *in* $U_a(V)$. *Moreover, the functional determinant of these* n *coordinate function is* $\neq 0$.

Conversely, given on a separable Hausdorff space M a countable open covering $\{U_\alpha\}$ and a homeomorphism $u_\alpha : U_\alpha \to E^n$ for each α, such that the condition 2) is satisfied. There existing a uniquely determined minimal differentiable structure of class k on M which admits the coordinate functions in $u_\alpha(U_\alpha)$ as functions of class k.

This theorem identifies our definition with the usual definition by means of "overlapping neighbourhoods". Its proof does not offer any difficulty, and we leave it to the reader. The disadvantage of the usual definition is that two families of neighbourhood may define the same differentiable structure, thus making the introduction of an equivalence relation necessary.

Definition 5.1.8. When a differentiable manifold $\{M, E_k\}$ has a covering $\{U_\alpha\}$ with the properties 1), 2) of Theorem 5.1.7, the sets U_α are called *coordinate neighbourhoods*. Relative to U_α (and to u_α) the coordinates of $u_\alpha(p)$ for $p \in U$ are called the *local coordinates* of p.

Definition 5.1.9. Given two differentiable manifolds $\{M, F_k\}$, $\{M', F_k'\}$, and a contінous mapping $\varphi : M \to M'$. To a function f'/U' in F_k' corresponds a function $f' \circ \varphi|\varphi^{-1}(U')$ defined in the open subset $\varphi^{-1}(U')$. The mapping φ is called *differentiable*, if the latter functions be long to F_k for all f' of F_k'. The two differentiable manifolds are called *isomorphic*, if φ is a homeomorphism and if both φ and φ^{-1} are differentiable.

The differentiable structures defined on the Euclidean line E^1 in Examples 5.1.4 and 5.1.6 above are isomorphic. It is not known whether the same space M can face the underlying space of two non-isomorphic differentiable manifolds.

Definition 5.1.10. Let $\{M, F_k\}$ and $\{M', F_k'\}$ be two differentiable manifolds of the same class k, let

$$\pi : M \times M' \to M, \qquad \pi' : M \times M' \to M'$$

be the projections defined respectively by

$$\pi(p, p') = p, \qquad \pi'(p, p') = p'.$$

To a function $f|U$ in F_k corresponds a function $f \circ \pi|\pi^{-1}(U)$. Similarly, to a function $f'|U'$ in F_k' corresponds a function $f' \circ \pi'|\pi'^{-1}(U')$. The family of these functions for all f in F_k and all f in F_k' has the property (P), and hence defines a differentiable structure of class k with the underlying space $M \times M'$. It will be called the product of the two manifolds, and will be denoted by $\{M \times M', F_k \times F_k'\}$.

5.2 Tangent space

One of the main consequences of the existence of a differentiable structure on a topological space is the possibility of "approximating" the space in the neighbourhood of a point by a linear space. In exact terms this is formulated by the introduction of the tangent space.

Considers differentiable manifold $\{M, F_k\}$, and a point $p \in M$. By a *function of class k at p* we mean a function of F_k with a domain of definition which is a neighbourhood of p. To such functions we introduce an equivalence relation: two functions f and g of class k at p belong to the same class if they are identical in a neighbourhood of p. This is clearly equivalence relation. We define addition of two classes of functions $\{f\}$ and $\{g\}$ by adding their representatives, and scalar multiplication of a class by multiplying its representative. With this definition of addition and scalar multiplication all classes of functions of class k at p form an infinite-dimensional vector space, to be denoted by S_p.

Let U be a neighbourhood of p, and $u : U \to u(U)$ a homeomorphism of U onto an open subset of E^n, defined by

$$u(q) = (u^1(q), \ldots, u^n(q)), \qquad q \in U,$$

where $u^1(q), \ldots, u^n(q)$ are n functions of class k in U. A function of class k at p being of the form $f = g \circ u$, with g of class $k \geq 1$, we say that *f has zero differential at p*, if all the first partial derivatives of g with respect to $u^1, \ldots, u^n$ vanish at $u(p)$. This property is obviously a property of a class of functions $\{f\}$ at p. It is also independent of the choice of the local coordinates $u^1, \ldots, u^n$. For, if $u'^1, \ldots, u'^n$ form another local coordinate system at p, the two representatives g, g' of f are related by the formula (6). Using the formula for the partial derivatives of a composite function, we get

$$\left(\frac{\partial g}{\partial u^i}\right)_{u(p)} = \sum_{j=1}^{n} \left(\frac{\partial g'}{du'^j}\right)_{u'(p)} \left(\frac{\partial u'^j}{du^i}\right)_{u(p)} \qquad i = 1, \ldots, n.$$

It follows that the vanishing of the partial derivatives $(\frac{\partial g'}{\partial u'^i})_{u'(p)}$ implies the vanishing of $(\frac{\partial g}{\partial u^i})_{u(p)}$.

Thus it is perfectly meaningful to speak of the classes of functions of zero differential at p. Clearly they form a linear subspace of S_p, which we shall denote by Z_p.

Definition 5.2.1. The quotient space $V_p^* = S_p/Z_p$ is called the space of tangent covectors at p, its elements being tangent covectors or covectors. The dual space V_p of V_p^* is called the space of tangent vectors or the tangent space at p, its elements being tangent vectors or vectors.

A covector at p is therefore a residue class relative to Z_p of a class of functions $\{f\}$. It is uniquely determined by a representative f, and we shall denote it by

$df(p)$ or df. On the other hand, a vector will be denoted by $X(p)$ or X. For their scalar product we shall adopt alternatively the notations:

$$\langle X(p), df(p)\rangle = \langle X, df\rangle = Xf. \tag{7}$$

Theorem 5.2.2. *Let $f, f^1, \dots, f^m$ be functions of class k at p, such that*

$$f = F(f^1, \dots, f^m), \tag{8}$$

where $F(f^1, \dots, f^m)$ is a function of class 1 in a neighbourhood of $(f^1(p), \dots, f^m(p))$. Then

$$df(p) = \sum_{j=1}^{m} \left(\frac{\partial F}{\partial f^j}\right)_{(f^1(p),\dots,f^m(p))} df^j(p). \tag{9}$$

For simplicity let us write

$$\alpha_j = \left(\frac{\partial F}{\partial f^j}\right)_{(f^1(p),\dots,f^m(p))}, \qquad j = 1, \dots, m.$$

The theorem then asserts that the function

$$f - \sum_{j=1}^{m} \alpha_j f^j$$

has zero differential at p. In terms of a local coordinate system $u^1, \dots, u^n$ at p let $g, g^1, \dots, g^m$ be the representations of $f, f^1, \dots, f^m$ respectively, so that

$$f = g \circ u, \qquad f^j = g^j \circ u.$$

Then we have, by applying the homeomorphism u^{-1} to (8),

$$g(u^1, \dots, u^n) = F(g^1(u^1, \dots, u^n), \dots, g^m(u^1, \dots, u^n)).$$

Hence the function

$$g - \sum_{j=1}^{m} \alpha_j g^j.$$

has all first partial derivatives equal to 0 at $u(p)$.

Corollary 5.2.3. *The covectors at p satisfy the identities:*

$$\begin{aligned} d(\alpha f + \beta g) &= \alpha df + \beta dg, \\ d(fg) &= f(p) dg + g(p) df, \end{aligned} \tag{10}$$

where f, g are functions of class k at p, and α, β are real numbers.

Corollary 5.2.4. *Every tangent space of a differentiable manifold of dimension n is a vector space of dimension n.*

Since every function f of class k at p is of the form $F(u^1,\dots,u^n)$, df is a linear combination of $du^1,\dots,du^n$. The latter are moreover linearly independent. For a relation of the form

$$\alpha_1 du^1 + \cdots + \alpha_n du^n = 0,$$

where $\alpha_1,\dots,\alpha_n$ are real numbers, implies, by Corollary 1,

$$d(\alpha_1 u^1 + \cdots + \alpha_n u^n) = 0,$$

which means that $\alpha_1 u^1 + \cdots + \alpha_n u^n$ has zero differential at p. But this is possible only when all the α's are zero. It follows that V_p^*, and hence V_p, is of dimension n.

Corollary 5.2.5.

$$\begin{aligned} X(\alpha f + \beta g) &= \alpha X f + \beta X g, \\ X(fg) &= f(p)Xg + g(p)Xf. \end{aligned} \tag{11}$$

Corollary 5.2.6. *If $u^1,\dots,u^n$ form a local coordinate system at p, $du^1,\dots,du^n$ form a base of V_p^*. Denote by $X_1,\dots,X_n$ its dual base in V_p. Then*

$$X_i f(p) = \left(\frac{\partial g}{\partial u^i}\right)_{u(p)}, \qquad f = g \circ u. \tag{12}$$

For by definition,

$$\langle X_i, du^j\rangle = \delta_i^j, \qquad i,j = 1,\dots,n.$$

It follows that

$$\begin{aligned} X_i f(p) = \langle X_i(p), df(p)\rangle &= \left\langle X_i(p), \sum_j \left(\frac{\partial g}{\partial u^j}\right)_{u(p)} du^j \right\rangle \\ &= \left(\frac{\partial g}{\partial u^j}\right)_{u(p)}. \end{aligned}$$

There are other ways of introducing the tangent space. One which is geometrically suggestive is by means of *parameterized curves.* We first observe from a manifold $\{M, F_k\}$, we can define, to an open subset U of M, a manifold with U as the underlying space by taking as its functions of class k the restrictions of the functions of F_k to open subsets of U. This differentiable structure on U is called an *induced structure*. In particular, there is an induced structure of class k on any open subset of the Euclidean line E^1. A *parametrized curve* in $\{M, F_k\}$ is a differentiable mapping of class k of an open subset of E^1 into M.

Let $p \in \mu$. Consider those parametrized curves whose open subsets in E^1 contain the origin 0 of E^1 and wich map 0 into p. Two such parametrized curves are called equivalent, if, for every function f as class k at p

$$\left(\frac{d(f\circ\lambda)}{dt}\right)_{t=0} = \left(\frac{d(f\circ\lambda')}{dt}\right)_{t=0}, \tag{13}$$

where λ and λ' are the mappings defining the parametrized curves and t is the coordinate on E^1. It is possible to define addition and scalar multiplication of such equivalence classes of parametrized curves at p. They then form a vector space which is naturally isomorphic to V_p. We leave the details of these considerations to the reader.

Consider now two manifolds $\{M, F_k\}$, $\{M', F_k'\}$ and a differentiable mapping of the first into the second defined by the continous mapping $\varphi : M \to M'$. Let $p \in M$, and let $p' = \varphi(p)$. To a function f' of class k at p' corresponds then the function $f' \circ \varphi$ of class k at p. This mapping induces a mapping of classes of functions and a linear map $S_{p'} \to S_p$. It maps $Z_{p'}$ into Z_p. In fact, let $u^1, \ldots, u^n$ and $u'^1, \ldots, u'^m$ be local coordinates at p and p' respectively, with the homeomorphisms u and u'.We can write the map pings under consideration in the sequence:

$$u(U) \xrightarrow{u^{-1}} U \xrightarrow{\varphi} U' \xrightarrow{u'} u'(U') \xrightarrow{g'} R,$$

where R is the real field and g' a function of class k in $u'(U')$. The condition that the mapping is differentiable implies that $g' \circ u' \circ \varphi \circ u^{-1}$ is of class k in $u(U)$. It follows that the functions

$$f = g' \circ u' \circ \varphi, \qquad f^j = u'^j \circ \varphi, \qquad j = 1, \ldots, m$$

are of class k in U and that we have, in a neighbourhood of $u'(p')$,

$$f = g'(f', \ldots, f^m).$$

If $g' \circ u'$ has zero differential at p', the partial derivatives $(\frac{\partial g'}{\partial f'^j})_{u'(p')}$ are zero. By Theorem 5.2.2, we get

$$df(p) = 0.$$

This proves that the mapping which sends f' into $f' \circ \varphi$ induces a mapping of $Z_{p'}$, into Z_p. It therefore induces a linear map $\varphi^* : V^*(p') \to V^*(p)$. The dual map of the latter will be denoted by $\varphi_* : V(p) \to V(p')$.

From the definition of the map φ^* we have the relation

$$\varphi^*(df'(p')) = d(f' \circ \varphi)(p), \tag{14}$$

where f' is a function of class k at p'. In particular, we have

$$\varphi^*(du'^j(p')) = d(u'^j \circ \varphi)(p).$$

Let the mapping $u' \circ \varphi \circ u^{-1}$ be defined by

$$u'^j = u'^j(u^1, \ldots, u^n).$$

where u'^j are functions of class k in $u(U)$. Introduce also the constants

$$\alpha_i^j = \left(\frac{\partial u'^j}{\partial u^i}\right)_{u(p)}. \tag{15}$$

Then we have, by applying Theorem 5.2.2.

$$\varphi^*(du'^j(p')) = \sum_{i=1}^{n} \alpha_i^j du^i(p). \tag{16}$$

Thus we see that to a differentiable mapping φ of one manifold into another corresponds a linear map φ_* of the tangent space $V(p)$ into the tangent space $V(p')$ at the image point $p' = \varphi(p)$. This linear map and its dual map of the spaces of covectors determine each other. The significance of these linear maps lies in the fact that their knowledge gives information on the mapping φ itself. In fact, we have the theorem.

Theorem 5.2.7. *Let φ define a differentiable mapping of $\{M, F_k\}$ into $\{M', F'_k\}$. Let $p \in M$, and $p' = \varphi(p)$. Then we have:*

1) *If $\varphi_* : V(p) \longrightarrow V(p')$is onto, there exists, to any system of local coordinates $u'^1, \ldots, u'^m$ at p', $n - m$ functions $u^{m+1}, \ldots, u^n$ at p, such that $u'^1 \circ \varphi, \ldots, u'^m \circ \varphi, u^{m+1}, \ldots, u^n$ form a system of local coordinates at p.*
2) *If φ_* is univalent, it is possible to select from $u'^1 \circ \varphi, \ldots, u'^m \circ \varphi$ n functions which form a system local coordinates at p. Moreover, if $u^1, \ldots, u^n$ form a system of local coordinates at p, there exist local coordinates $u'^1, \ldots, u'^m$ at p' such that $u^i = u'^i \circ \varphi$, $i = 1, \ldots, n$, in a neighbourhood of p.*
3) *If φ_* is a one-one isomorphism, φ defines a homeomorphism of a neighbourhood of p onto a neighbourhood of p', such that its inverse mapping is differentiable at p'.*

Remark 5.2.8. The converses of these statements are not true. Take the structures $\{x\}$ and $\{x^3\}$ on the Euclidean line and consider the mapping $x \longrightarrow x^3$ at O.

The proof of this theorem can be found in Chevalley, Lie Groups, pp. 79-80, and is therefore omitted.

5.3 Tensor bundles

From the space $V(p)$ or $V^*(p)$ at a point p of a manifold we can construct the space $V_s^r(p)$ of tensors of type (r,s), the space of exterior r-vectors $\wedge^r(V)(p)$, etc., according to the results Chapter 4. Take for definiteness the spaces $V_s^r(p)$, and consider the set

$$T_s^r = \bigcup_{p \in U} V_s^r(p). \tag{17}$$

We shall define on it a topology, making in into a separable Hausdorff space, and a differentiable structure of class $k-1$ ($k-1=\infty$, if $k=\infty$). Moreover, relations between T_s^r and M will be established, which will make T_s^r into a fiber bundle over M.

Let $\tilde{V}$ be a vector space of dimension n over the real field, and $\tilde{V}_s^r$ the corresponding space of tensors of type (r,s). We fix a base in $\tilde{V}$, It gives rise to a base in $\tilde{V}_s^r$, so that an element in $\tilde{V}_s^r$ is determined by its components, and vice versa.

Consider now a coordinate neighbourhood U in M, with the homeomorphism $u: U \to u(U) \subset E^n$. For a point $p \in U$ denote its local coordinates by $u^1(p), \ldots, u^n(p)$. This local coordinate system defines a base $du^1(p), \ldots, du^n(p)$ in $V^*(p)$, a dual base $X_1(p), \ldots, X_n(p)$ in $V(p)$, and hence a well-defined base in $V_s^r(p)$ we shall call them the *natural bases*, which are completely determined by a local coordinate system. We now define the mapping

$$\varphi_U : U \times V_s^r \to \bigcup_{p \in U} V_s^r(p) \tag{18}$$

by the condition that $\varphi_U(p,y)$, for $p \in U$, $y \in V_s^r$ belongs to $V_s^r(p)$ and has the same components relative to the natural base as the components of y in $\tilde{V}_s^r$. This is clearly a one-one mapping.

Suppose V be a second coordinate neighbourhood with $U \cap V \neq 0$. We define the mapping $\varphi_{U,p} : \tilde{V}_s^r \to V_s^r(p)$ by

$$\varphi_{U,p}(y) = \varphi_U(p,y), \qquad y \in \tilde{V}_s^r. \tag{19}$$

Then

$$g_{UV}(p) = \varphi_{U,p}^{-1} \circ \varphi_{V,p} \tag{20}$$

is a one-one mapping of $\tilde{V}_s^r$ onto itself, whose equations we proceed to determine. For simplicity of writing we restrict ourselves to the case $r = s = 1$. The local coordinates in V being $v^1, \ldots, v^n$, we denote the natural bases by $dv^1(p), \ldots, dv^n(p)$, and $Y_1(p), \ldots, Y_n(p)$. Put

$$y' = g_{UV}(p)y. \tag{21}$$

Then equation (20) gives

$$\varphi_U(p,y') = \varphi_V(p,y). \tag{22}$$

If the components of y and y' in $\tilde{V}_1^1$ are denoted by y^i_j and y'^i_j respectively, we have by definition,

$$\begin{aligned} \varphi_U(p,y') &= \sum_{i,j} y'^i_j X_i(p) \otimes du^j(p), \\ \varphi_V(p,y) &= \sum_{i,j} y^i_j Y_i(p) \otimes dv^j(p). \end{aligned} \tag{23}$$

On the other hand, it follows from Theorem 5.2.2 that

$$dv^j(p) = \sum_i \left(\frac{\partial v^j}{\partial u^i}\right)_{u(p)} du^i(p) \tag{24}$$

and hence that

$$Y_j(p) = \sum_i \left(\frac{\partial u^i}{\partial v^j}\right)_{v(p)} X_i(p). \tag{24'}$$

From (22),(23),(24), (24′) we get

$$y'^i_j = \sum_{\ell,m} y^\ell_m \left(\frac{\partial u^i}{\partial v^\ell}\right)_{v(p)} \left(\frac{\partial v^m}{\partial u^j}\right)_{u(p)}. \tag{25}$$

These equations define $g_{UV}(p)$, being therefore a linear auto-morphism in $\tilde{V}_1^1$. We give to $\tilde{V}_1^1$ the topology and differentiable structure of class ∞ derived from the Euclidean space of the components of its elements. Let $GL(\tilde{V}_1^1)$ be the general linear (topological) group in $\tilde{V}_1^1$. Then $g_{UV}(p)$ defines a Mapping $g_{UV} : U \cap V \to GL(\tilde{V}_1^1)$. Since the manifold M is of class $k \geq 1$, we derive from (25) the important fact that the mappings g_{UV} are continous in $U \cap V$.

We take a covering of M by the coordinate neighbourhood U, V, W, etc., and denote the corresponding mappings (18) by φ_U, φ_V, φ_W, etc. Since every mapping g_{UV} is continuous, it follows from the theory of fiber bundles that T_1^1 is a fiber bundle over M with φ_U as the coordinate functions. (Cf. Steenrod, Fiber Bundles, p. 14). This defines meanwhile a topology in T_1^1 characterized by the condition that φ_U maps open sets of $U \cap \tilde{V}_1^1$ into open sets of T_1^1. We leave the reader to verify that this topology makes T_1^1 into a separable Hausdorff space.

The space T_1^1 has the open covering

$$\{\varphi_U(U \times \tilde{V}_1^1), \varphi_V(V \times \tilde{V}_1^1), \ldots\}.$$

To every member of the covering, say, $\varphi_U(U \times \tilde{V}_1^1)$, $u^i(p)$, y^i_j are the local coordinates. Since the functions in the right-hand side of (25) are of class $k-1$, this covering defines on T_1^1 a differentiable structure of class $k-1$.

It is hardly necessary to remark that these considerations can be carried out for the tensor bundles T_s^r, the bundles of exterior forms $\wedge^r(M^*)$, etc.

Consider again the tensor bundle T_s^r. Denote by

$$\psi : T_s^r \to M$$

the projection defined by $\psi(V_s^T(p)) = p$. For $\ell < k$, a *cross-section of class* ℓ is a differentiable mapping $f : M \to T_s^r$ of class ℓ such that $\psi f =$ identity, Such a cross section is also called a *tensor field of class* ℓ *and type* (r,s). In particular, a tensor field of type $(1,0)$ is called a *vector field* or an *infinitesimal transformation.* A cross section in the bundle of exterior forms is called an exterior *differential form.*

The question on the existence of cross sections with certain properties (for instance, a vector field such that the vectors are nowhere zero) is of great importance and interest in the theory of differentiable manifolds. Concerning this question differentiability requirements do not complicate the problem, for it can be proved that every continuous cross section can be approximated by a differentiable cross section. (Cf. Steenrod, pp. 25–28).

5.4 Submanifolds; Imbedding of compact manifolds

Definition 5.4.1. A submanifold of a differentiable manifold M of class k consists of a manifold N of class k and a differentiable mapping $\varphi : N \to M$ such that 1), φ is one-one; 2) For every $q \in N$ the induced linear map $\varphi_* : V(q) \to V(\varphi(q))$ on the tan gen spaces is univalent. We also say that the submanifold (φ, N) is imbedded in M.

Condition 1) implies that φ is a continuous mapping. It is not necessarily a homeomorphism (into), as shown by some examples given below.

Condition 2) implies that $\dim N \le \dim M$.

Example 5.4.2. Open submanifold of a manifold. Here N is an open subset of M, with the differentiable structure of class k defined by the family of functions which are restrictions of functions of class k on M to open subsets of N. φ is the identity mapping.

Example 5.4.3. Closed submanifold of a manifold. It is a submanifold (φ, N) satisfying the conditions: 1) $\varphi(N)$ is a closed subset of M; 2) Every point $p \in \varphi(N)$ is contained in a coordinate neighbourhood U with the local coordinates $u^1, \dots, u^n$ such that the set $\varphi(N) \cap U$ is a coordinate slice defined by the equations $u^{r+1} = 0, \dots, u^n = 0$.

Those are the submanifolds which form the object of study in classical differential geometry.

Example 5.4.4. The twisted torus let the torus M be defined by coordinates (x, y), where x, y are real number mod 1. Take the real line E^1 with coordinate t, $-\infty < t < +\infty$. Let a, b be a pair of real numbers whose ratio is irrational. Define $\varphi(t) = (at, bt)$. Then (φ, E^1) is a submanifold of M. The set $\varphi(E^1)$ is everywhere dense in M, so that φ is not a homeomorphism.

Example 5.4.5. Here are some examples of imbedding the Euclidean line into the Euclidean plane. The definitions are clear from the figures.

Guided by the fact that a plane curve which crosses itself has a rather harmless singularity at the self-intersection we are lead to the definition:

Definition 5.4.6. A manifold N of class k is said to be *immersed* in a manifold M of class k if there is a differentiable mapping $\varphi : N \to M$ such that:

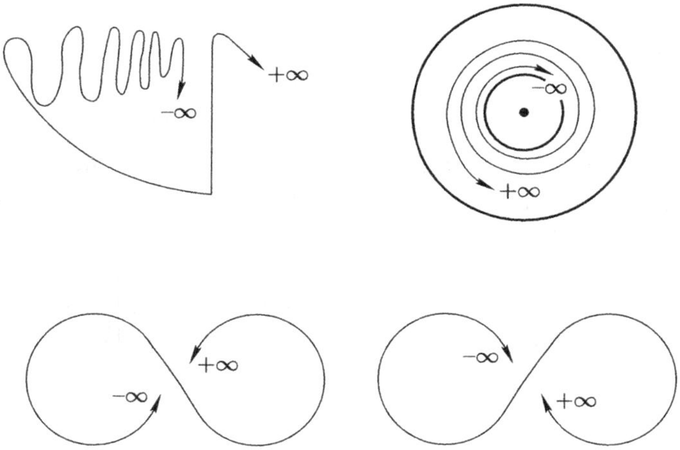

Fig. 1.

1) For every $q \in N$ the induced linear map $\varphi_* : V(q) \to V(\varphi(q))$ on the tangent spaces is univalent;
2) If $\varphi(q) = \varphi(q'), q \neq q'$, the spaces $\varphi_*(V(q))$ and $\varphi_*(V(q'))$ have only the zero vector in common.

Definition 5.4.7. The submanifold (φ, N) is said to be *regularly imbedded* in M if φ is a homeomorphism (into).

To have a more geometrical picture of the notion of regular imbedding we introduce the *limit set* $L(\varphi)$ of a submanifold. It consists of all points $p \in M$ such that is a divergent sequence q_i in N whose image points $\varphi(q_i)$ converge to p. It is easy to prove that (φ, N) is regularly imbedded if and only if $\varphi(N) \cap L(\varphi) = 0$.

Theorem 5.4.8. *A submanifold (φ, N) is regularly imbedded in M, if and only if it is a closed submanifold of an open submanifold of M. Moreover, the subset $\varphi(N)$ has then a differentiable structure of class k which is isomorphic to any submanifold regularly imbedded in M and having $\varphi(N)$ as the image set.*

In other words, if we restrict ourselves to regular imbedding, the structure of the submanifold is "carried" by the image set. Unfortunately, geometrical problem, and in particular problems concerning the integration of differential systems, have shown the necessity of considering submanifolds no regularly imbedded.

To prove the theorem we first observe that the sufficiency is trivial, that is, every closed submanifold of an open submanifold of M is regularly imbedded in M. Suppose now the submanifold be regularly imbedded. Let $q \in N$, $p = \varphi(q)$. Denote by n, r the dimensions of M, N respectively. There are coordinates $u^1, \ldots, u^n$ at p such that $u^1 \circ \varphi, \ldots, u^r \circ \varphi$ form a coordinate system at q. The remaining $n - r$ functions $u^\lambda \circ \varphi$, $\lambda = r+1, \ldots, n$, are then functions of class k in these coordinates at q:

$$u^\lambda \circ \varphi = F^\lambda(u^1 \circ \varphi, \ldots, u^r \circ \varphi).$$

Replacing the coordinates u^λ by $u^\lambda - F^\lambda(u^1,\ldots,u^r)$ and denoting them again by u^λ, we have $u^\lambda \circ \varphi = 0$. Since φ is a homeomorphism, there is a neighbourhood u_p of p such that $\varphi(N) \cap U_p$ is the coordinate slice defined by $u^\lambda = 0$. Let $V = \bigcap_{p\in\varphi(N)} U_p$. Then V is an open submanifold of M which contains $\varphi(N)$. The set $\varphi(N)$ is relatively closed in V. For if $p' \in V \cap \overline{\varphi(N)}$, p' belongs to a certais U_p. Since $\varphi(N) \cap U_p$ is relatively closed in U_p, $p' \in \varphi(N)$. It is clear from the definition of V that (φ, N) is a closed submanifold of V.

The functions $u^1|\varphi(N) \cap U, \ldots, u^r|\varphi(N) \cap U$ define a local coordinate system on $\varphi(N)$, and hence a differentiable structure on $\varphi(N)$. The latter is isomorphic to N. Since its definition does not depend on N, we have proved the second statement in the theorem.

Theorem 5.4.9. *A compact differentiable manifold of class k can be imbedded in a Euclidean space of sufficiently high dimension.*

Let U_j, $j = 1,\ldots,s$ be a finite covering of the manifold M by coordinate neighbourhoods such that there exist coordinate neighbourhoods V_j, with $\overline{U}_j \subset V_j$. The manifold M being of dimension n, we denote the local coordinates of a point $p \in V_j$ by $v_j^1(p),\ldots,v_j^n(p)$. To each j let $f_j(p)$, $p \in M$, be a function of class k such that $0 \le f_j(p) \le 1$, and that

$$\begin{aligned} f_j(p) &= 1, \quad p \in \overline{U}_j,\\ f_j(p) &= 0, \quad p \in M - V_j. \end{aligned} \tag{26}$$

We now define the $(n+1)s$ functions of class k on M:

$$x_j^0(p) = f_j(p),$$

$$x_j^i(p) = \begin{cases} v_j^i(p) f_j(p), & p \in V_j \\ 0, & p \in M - V_j \end{cases} \quad i = 1,\ldots,n. \tag{27}$$

Regarding these functions as the coordinates of a point in a Euclidean space of dimension $(n+1)s$, we get a mapping $M \to E^N$, where $N = (n+1)s$. This mapping is clearly continuous. The image set is therefore compact, and hence closed. It is also one-one. For, suppose

$$\begin{aligned} x_j^0(p) &= x_j^0(q),\\ x_j^i(p) &= x_j^i(q). \end{aligned}$$

There must exist a j with

$$x_j^0(p) = 1.$$

Then

$$x_j^0(q) = 1,$$

which implies that $q \in V_j$. Moreover, we have

$$v_j^i(p) = v_j^i(q),$$

so that p, q are in the same coordinate neighbourhood and have the same local coordinates. This is possible only when $p = q$. Finally it is a simple matter to verify that this mapping satisfies condition (2) of Definition 5.4.1 and hence realizes an imbedding of M in E^N.

This imbedding theorem has important consequences. Whitney proved that any differentiable manifold of dimension n can be imbedded in a Euclidean space of dimension $2n+1$. The proof of Whitney's theorem is, however, more difficult.

6

Exterior Differential Forms

6.1 Exterior differentiation

Let M be a differentiable manifold of class $k \geq 1$. For a point $p \in M$ denote by $\wedge^*(p)$ the algebra of exterior forms of the tangent space at p. According to the results of the last Chapter the space

$$\wedge^*(M) = \bigcup_{p \in M} \wedge^*(p) \tag{1}$$

is a fiber bundle over M, and is a differentiable manifold of class $k-1$. An *exterior polynomial* of class $\ell \leq k-1$ is a cross section ω of class ℓ of this bundle. We have, therefore, $w(p) \in \wedge^*(p)$. If, for all $p \in M$, $\omega(p) \in \wedge^{*r}(p)$ for a fixed r, the exterior polynomial is an *exterior differentiable form of degree r*, or simply a *form* of degree r.

We define addition and (exterior) multiplication of exterior polynomials according to the formulas:

$$\begin{aligned}(\omega + \omega')(p) &= \omega(p) + \omega'(p), \\ (\omega \wedge \omega')(p) &= \omega(p) \wedge \omega'(p).\end{aligned} \tag{2}$$

It is clear that all exterior polynomials of class ℓ form a ring.

Theorem 6.1.1. *Let M be a differentiable manifold of class $k \geq 2$. There is one, and only one, mapping d, called exterior differentiation, which sends exterior polynomials of class ℓ (≥ 1) into exterior polynomials of class $\ell - 1$ and exterior differential forms of degree r into those of degree $r+1$, such that the following conditions are satisfied:*

1) $d(\omega + \omega') = d\omega + d\omega'$
2) *If ω is an exterior differential form of degree r,*

$$d(\omega \wedge \omega') = d\omega \wedge \omega' + (-1)^r \omega \wedge d\omega'. \tag{3}$$

3) *If $\omega = f$ is of degree O and is therefore a real-valued function, df is the same as the covector in the sense of* §5.2.

$$d(df) = 0. \tag{4}$$

4) *If f if a real-valued function of class $\geqq 2$, then*

It is obviously sufficient to prove the theorem by using-a local coordinate system $u^1, \dots, u^n$. In the coordinate neighbourhood U in question a form can be written

$$\omega = \sum_{i_1 < \cdots < i_r} a_{i_1 \dots i_r}(u^1, \dots, u^n) du^{i_1} \wedge \cdots \wedge du^{i_r}, \quad i_1, \dots, i_r = 1, \dots, n \tag{5}$$

while an exterior polynomial is a sum of such forms. By l) and 2) it is sufficient to define d as applied to f and df. But these are given by 3) and 4). This proves the uniqueness of the operation d.

The existence is proved by an explicit formula. In fact, for the form in 5, we define

$$d\omega = \sum_{i_1 < \cdots < i_r} da_{i_1 \dots i_r} du^{i_1} \wedge \cdots \wedge du^{i_r}. \tag{6}$$

It is easy to verify that the conditions l) - 4) are fulfilled.

Theorem 6.1.2. *For any exterior polynomial w of class ≥ 2,*

$$d(d\omega) = 0. \tag{7}$$

Since d fulfills condition 1) of the last theorem, it suffices to establish (7) for the case that ω is a "monomial"

$$\omega = a(u^i) du^{i_1} \wedge \cdots \wedge du^{i_r} \tag{8}$$

Then we have

$$d\omega = da \wedge du^{i_1} \wedge \cdots \wedge du^{i_r},$$

from which (7) follows.

Let M and M' be two manifolds of class k, and $\varphi : M \to M'$ be a differentiable mapping. If $\varphi(p) = p'$, there is an induced linear map $\varphi^* : \wedge^*(p') \to \wedge^*(p)$. To an exterior polynomial ω' in M' corresponds an exterior polynomial $\varphi^*\omega'$ in M according to the definition

$$\varphi^*\omega'(p) = \varphi^*(\omega'(p')). \tag{9}$$

If ω' is a form of degree r, $\varphi^*\omega'$ is also a form of degree r. We call φ^*, as applied on exterior polynomials, the dual mapping of φ.

Theorem 6.1.3. *Exterior differentiation commutes with the dual mapping of exterior polynomials under a differentiable mapping; that is,*

$$\varphi^* d\omega' = d\varphi^*\omega'. \tag{10}$$

It is sufficient to prove (10) for the case that ω' is a monomial. We then proceed by induction on its degree. When the degree is 1, we can write

$$\omega' = adv,$$

where v is one of the local coordinates in M'. Then, we have

$$\begin{aligned}\varphi^* d\omega' &= \varphi^*(da \wedge dv) = (\varphi^* da) \wedge (\varphi^* dv) \\ &= d(\varphi^* a) \wedge d(\varphi^* v) = d((\varphi^* a) d(\varphi^* v)) = d\varphi^*(adv),\end{aligned}$$

so that (10) is true. Suppose the formula is valid for monomials of degree r and ω' is a monomial of degree $r+1$. We write

$$\omega' = \omega_1' \wedge \omega_2',$$

where ω_1', ω_2' are forms of degrees 1 and r respectively. Applying the induction hypothesis, we have

$$\begin{aligned}\varphi^* d\omega' &= \varphi^*(d\omega_1' \wedge \omega_2' - \omega_1' \wedge dw_2') = \varphi^* d\omega_1' \wedge \varphi^* \omega_2' - \varphi^* \omega_1' \wedge \varphi^* dw_2' \\ &= d\varphi^* \omega_1' \wedge \varphi^* \omega_2' - \varphi^* \omega_1' \wedge d\varphi^* \omega_2' = d(\varphi^* \omega_1' \wedge \varphi^* \omega_2') = d\varphi^* \omega'\end{aligned}$$

This completes the induction, and the theorem is proved.

6.2 Differential systems; Frobenius's theorem

Consider a real vector space V of dimension n. By Theorem 4.3.11, Chap. 4, the linear subspaces of dimension $r(\leq n)$ of V are in one-one correspondence with the classes of non-zero decomposable r-vectors, each class consisting r-vectors differing from each other by a scalar factor. The set or r-vectors can be given a topology by means of the components in terms of a base. This gives rise to a topology and in fact to a differentiable structure in the set of linear subspaces of dimension r in V. We denote the manifold so obtained by $G^r(V)$ and call it a *Grassmann manifold.*

Similarly, we can define the Grassmann manifold $G^r(V^*)$ of the dual space V^* of V. By the choice of a base in $\wedge^n(V^*)$ it follows from Theorem 4.4.5, that there is a one-one mapping

$$D : G^r(V) \to G^{n-r}(V^*). \tag{11}$$

Since a base in $\wedge^n(V^*)$ is defined up to a factor, this mapping ia independent of the choice of the base. It is clearly a homeomorphism. Geometrically it maps a linear subspace of dimension in V into its orthogonal linear subspace of dimension $n-r$ in V^*.

As in the case of tensor bundles, the set

$$G^r(M) = \bigcup_{p \in M} G^r(V(p)), \tag{12}$$

where M is a manifold of class k, can be given a topology and a differentiable structure of class $k-1$, and is a fiber bundle over M. Similarly,

$$G^{*n-r}(M) = \bigcup_{p \in M} G^{n-r}(V^*(p)) \tag{13}$$

is a fiber bundle over M.

Definition 6.2.1. A cross section $F : M \to G^r(M)$ of the bundle (12) is called a differential system of dimension r. Such a cross section is equivalent to one of the bundle $G^{*n-r}(M)$ over M.

A differential system of dimension r therefore associates to every point of M a linear subspace of dimension r of its tangent space, or, what is the same, its orthogonal linear subspace of dimension $n-r$ of the dual space. The latter defines, a decomposable form of degree $n-r$, determined up to a non-zero factor.

A manifold may not have differential systems of a given dimension. In fact, the existence of a differential system of dimension 1 is equivalent to that of a field of directions. It is well-known for instance that the latter is not possible on an even-dimensional sphere. However, in such cases we can generally omit the "singularities" and restrict ourselves to an open submanifold in which the differential system exists. The theorems to be established below will then retain their validity.

Definition 6.2.2. Let F be a differential system, so that, for any $p \in M$, $F(p)$ is a linear subspace of dimension r of the tangent space $V(p)$ at p. A submanifold (φ, N) is called an integral manifold of the differential system if, for any $q \in N$, $\varphi_*(V(q))$ is contained in $F(\varphi(q))$.

Since φ_* is univalent, it follows that an integral manifold is of dimension $\leq r$.

Definition 6.2.3. A differential system F of dimension r is called completely integrable if every point $p \in M$ has a coordinate neighbourhood with the local coordinates $u^1, \ldots, u^n$ such that the coordinate slices $u^{r+1} = \text{const.}, \ldots, u^n = \text{const.}$, are integral manifolds of F.

The main purpose of this section is to formulate a condition that a differential system F be completely integrable. Let $p \in M$ and U a coordinate neighbourhood containing p. If F is of class $\ell \geq 1$, it defines and can be defined in U, by a non-zero decomposable form Ω of degree $n-r$, determined up to a non-zero factor. The system F is said to satisfy the condition (C) at p, if U can be so chosen that $d\Omega$ is a multiple of Ω by a linear differential form, or, in other words, that $d\Omega$ belongs to the ideal generated by Ω in the ring of exterior polynomials in U. The condition (C) is clearly independent of the choice of the coordinate neighbourhood. It also remains invariant if we replace Ω by $\Omega^* = \lambda\Omega$, where λ is a non-zero function os class ℓ. For we have

$$d\Omega^* = \lambda d\Omega + d\lambda \wedge \Omega.$$

Therefore condition (C) is a property of the system F.

Theorem 6.2.4. *(Frobenius) A necessary and sufficient condition that a differential system F of class ≥ 1 be completely integrable is that the condition (C) be satisfied at all points $p \in M$.*

The condition is clearly necessary, for by using local coordinates $u^1, \ldots, u^n$ in Definition 6.2.3, we can take

$$\Omega = du^{r+1} \wedge \cdots \wedge du^n,$$

which fulfills condition (C).

To prove the sufficiency, we need the following lemma.

Lemma 6.2.5. *Suppose a differential system F of dimension r has the property that at every point $p \in M$ there is a local coordinate system $u^1, \ldots, u^n$ such that its corresponding differential form can be chosen to involve only $n - r + 1$ of these coordinates. Then the system is completely integrable.*

We suppose $u^i(p) = 0$ and suppose the coordinate neighbourhood to be defined by $|u^i| < \varepsilon$, $i = 1, \ldots, m$. Let u^λ, $\lambda = r, \ldots, n$, be the coordinates involved in the differential form Ω corresponding to the system F. Then we can write

$$\Omega = \sum_\lambda (-1)^\lambda a^\lambda du^r \wedge \cdots \wedge du^{\lambda-1} \wedge du^{\lambda+1} \wedge \cdots \wedge du^n, \tag{14}$$

where a^λ are functions of class ≥ 1 in u^μ, $\mu = r, \ldots, n$.

Consider now the $(n - r + 1)$-dimensional space of the coordinates u^λ, and denote by U_1 the neighbourhood defined by $|u^\lambda| < \varepsilon$. By Theorem 4.4.5, as applied to the tangent spaces at points of U_1, there is an isomorphism between the $(n-r)$-forms and the vector fields of U_1. To achieve this we have to choose a generator in $\wedge^{n-r+1}(V_1^*(q))$, where $V_1(q)$ is the space of covectors at a generic point $q \in U_1$. Since the vectors $\frac{\partial}{\partial u^\lambda}$ and the covectors du^μ form dual bases in the tangent spaces and their dual spaces, such a generator can be chosen to be $du^r \wedge \cdots \wedge du^n$. Recalling the definition of an interior product (Chap. 4, (59)), we find

$$\left\langle \frac{\partial}{\partial u^{\lambda_1}} \wedge \cdots \wedge \frac{\partial}{\partial u^{\lambda_{n-r}}}, \left(\sum_\lambda a^\lambda \frac{\partial}{\partial u^\lambda}\right) \rfloor (du^r \wedge \ldots \wedge du^n) \right\rangle$$

$$= \left\langle \frac{\partial}{\partial u^{\lambda_1}} \wedge \cdots \wedge u^{\lambda_{n-r}} \wedge \left(\sum_\lambda a^\lambda \frac{\partial}{\partial u^\lambda}\right), du^r \wedge \cdots \wedge du^n \right\rangle$$

$$= \operatorname{sgn}(\lambda_1, \ldots, \lambda_{n-r}, \lambda) a^\lambda,$$

where $\operatorname{sgn}(\lambda_1, \ldots, \lambda_{n-r}, \lambda) \neq 0$ only when the indices are all distinct and is then equal to $+1$ or -1 according as they form an even or an odd permutation of $r, \ldots, mn$. This proves that the vector orthogonal to Ω is, up to a sign, $x = \sum_\lambda \alpha^\lambda \frac{\partial}{\partial u^\lambda}$. The relation between Ω and x is reciprocal.

The functions a^λ are not all zero. We suppose $a^r(0,\dots,0) \neq 0$ and suppose U to be so small that $a^r \neq 0$ in U_1. Consider the system of ordinary differential equations

$$\frac{du^j}{du^r} = \frac{a^j}{a^r}, \quad j = r+1,\dots,n. \tag{15}$$

By a well-known existence theorem there is a solution

$$u^j = f^j(u^r, v^{r+1},\dots,v^n), \tag{16}$$

which takes, for $u^r = 0$ the preassigned initial values

$$v^j = f^j(0, v^{r+1},\dots,v^n), \tag{17}$$

provided that the latter are sufficiently small. Moreover, the functions f^j are of class 1 in v^k, $k = r+1,\dots,n$. Since the functional determinant. Since the functional determinant

$$\left|\frac{\partial f^j}{\partial v^k}\right| \neq 0$$

at $p_1 = (0,\dots,0)$, it follows from the theorem on implicit functions that the functions v^λ defined by

$$\begin{aligned} u^r &= v^r, \\ u^j &= f^j(v^r, v^{r+1},\dots,v^n), \end{aligned} \tag{18}$$

form a coordinate system at p_1. By (18) we get

$$\begin{aligned} du^j &= \frac{\partial f^j}{\partial v^r} dv^r + \sum_k \frac{\partial f^j}{\partial v^k} dv^k \\ &= \frac{a^j}{a^r} du^r + \sum_k \frac{\partial f^j}{\partial v^k} dv^k. \end{aligned}$$

Since we restrict ourselves to a neighbourhood in which the functional determinant $|\frac{\partial f^j}{\partial v^k}| \neq 0$, it follow that the system of differential equations (15) reduces to the system

$$\frac{dv^k}{dv^r} = 0$$

in the coordinates v^λ. This implies that, by the multiplication of Ω by a non-zero factor when necessary, we have

$$\Omega = dv^{r+1} \wedge \cdots \wedge dv^n. \tag{19}$$

Now Equations (18), together with

$$u^1 = v^1,\dots,u^{r-1} = v^{r-1}, \tag{20}$$

define a coordinate system v^i at p. Relative to v^i, Ω is given by (19), It follows that at every point $q \in U$ the linear space $F(q)$ defined by the differential system is spanned by the vectors $\frac{\partial}{\partial v^1}, \dots, \frac{\partial}{\partial v^r}$. Such a system is clearly completely integrable, so that the lemma is proved.

To prove the sufficiency of the condition in Theorem 6.2.4, we proceed by induction on the dimension r of the system F. For $r = 1$ this is contained in the lemma. Suppose the condition is sufficient for $r-1$. Since $\Omega \neq 0$, there is one of the differentials du^i say du^n, such that $\Omega \wedge du^n \neq 0$ in U (by reducing if necessary). The form $\Omega \wedge du^n$ defines in U a differential system F' of dimension $r-1$ which clearly satisfies condition (C). At any point $q \in U$ we have $F'(q) = F(q) \cap L(q)$, where $L(q)$ is the linear space of dimension $n-1$ spanned by $\frac{\partial}{\partial u^1}, \dots, \frac{\partial}{\partial u^{n-1}}$. To the system F' we can apply our induction assumption. It follows that a coordinate system $V^1, \dots, V^n$ exists at p such that $dv^r, \dots, dv^n$ are $n-r+1$ covectors orthogonal to $F'(q)$. As $n-r$ covectors orthogonal to $f(q)$ we can therefore take their linear combinations. Since Ω is the exterior product of $n-r$ linearly independent covectors orthogonal to $F(q)$, it follows that Ω is of the form

$$\Omega = \sum_{\lambda_1 < \cdots < \lambda_{n-r}} a_{\lambda_1 \dots \lambda_{n-r}} dv^{\lambda_1} \wedge \cdots \wedge dv^{\lambda_{n-r}}, \quad \lambda' s = r_1, \dots, n.$$

where $a_{\lambda_1 \cdots \lambda_{n-r}}$ are functions of class 1 in v^i. By hypothesis its exterior derivative is a multiple of Ω by a linear differential form, so that we have

$$\begin{aligned} d\Omega &= \sum_{i, \lambda_1 < \cdots < \lambda_{n-r}} \frac{\partial a_{\lambda_1 \cdots \lambda_{n-r}}}{\partial v^i} dv^i \wedge dv^{\lambda_1} \wedge \cdots \wedge dv^{\lambda_{n-r}} \\ &= (b_1 dv^1 + \cdots + b_n dv^n) \wedge \sum_{\lambda_1 < \cdots < \lambda_{n-r}} a_{\lambda_1 \dots \lambda_{n-r}} dv^{\lambda_1} \wedge \cdots \wedge dv^{\lambda_{n-r}}, \qquad (21) \\ i &= 1, \dots, n. \end{aligned}$$

Comparing the coefficients in both sides of terms involving dv^α, $\alpha = 1, \dots, r-1$, we get

$$\frac{\partial a_{\lambda_1 \cdots \lambda_{n-r}}}{\partial v^\alpha} = b_\alpha a_{\lambda_1 \dots \lambda_{n-r}}.$$

This implies, if a and a' denote two arbitrary coefficients in Ω,

$$a' \frac{\partial a}{\partial v^\alpha} - a \frac{\partial a'}{\partial v^\alpha} = 0.$$

It follows that the rations of the coefficients in Ω are Independent of v^α. Multiplying Ω by a factor when necessary, we can assume that Ω involves only v^λ. By the above lemma we conclude that the system F is completely integrable.

Corollary 6.2.6. *Let Ω be locally decomposed into a product of linear differential forms*

$$\Omega = \omega^{r+1} \wedge \cdots \wedge \omega^n, \tag{22}$$

so that ω^j, $j = r+1, \ldots, n$ *are linearly independent covectors orthogonal to* $F(q)$, $q \in U$. *The system* F *is completely integrable if and only if*

$$d\omega^j \wedge \omega^{r+1} \wedge \cdots \wedge \omega^n = 0, \tag{23}$$

that is, if and only if $d\omega^j$ *belongs to the ideal generated by* ω^k, $k = r+1, \ldots, n$.

In fact, we observe that the condition

$$d\Omega = \vartheta \wedge \Omega$$

is equivalent to

$$d\Omega \cap \omega^j = 0. \tag{24}$$

On the other hand, we have

$$d\Omega = \sum_j (-1)^{j+r-1} d\omega^j \wedge \omega^{r+1} \wedge \cdots \wedge \omega^{j-1} \wedge \omega^{j+1} \wedge \cdots \wedge \omega^n. \tag{25}$$

Equations (24) and (25) give (23).

It may be of interest to mention that Frobenius' Theorem generalizes well-known theorems on total differential equations. Consider the particular case $n = 3$, $r = 2$, Writing x, y, z for local coordinates, we have

$$\Omega = P(x,y,z)dx + Q(x,y,z)dy + R(x,y,z)dz, \tag{26}$$

from which we get

$$\begin{aligned} d\Omega \wedge \Omega =& \{(R_y - Q_z)dy \wedge dz + (P_z - R_x)dz \wedge dx + (Q_x - P_y)dx \wedge dy\} \\ & \wedge (Pdx + Qdy + Rd) \\ =& \{P(R_y - Q_z) + Q(P_z - R_x) + R(Q_x - P_y)\}dx \wedge dy \wedge dz. \end{aligned}$$

By the above corollary the condition for complete integrability is

$$P(R_y - Q_z) + Q(P_z - R_x) + R(Q_x - P_y) = 0. \tag{27}$$

Frobenius's Theorem is a "local" theorem in the sense that it deals with the behavior of a differential system in a neighbourhood. With this information it is, however, possible to study the integral manifolds in the large. Chevalley did this by introducing a new topology in M, We refer to his "Lie Groups" for details, pp. 92–94.

6.3 Derivations and anti-derivations

In dealing with the concepts and operations which have been and which will be developed, it seem desirable to introduce some general notions and establish their properties.

Definition 6.3.1. A graded ring A is a ring which has the following properties:

1) Its additive group is a graded group, that is, direct sum

$$A = A^0 + A^1 + A^2 + \cdots . \tag{28}$$

2) If $x \in A^k$, $y \in A^i$, then $xy \in A^{k+i}$. An element of A^k is said to be homogeneous of degree k.

Examples of graded rings:

1) The exterior algebra of a vector space V of dimension n:

$$\wedge(V) = \wedge^0(V) + \wedge^1(V) + \cdots + \wedge^n(V).$$

2) The tensors algebra of contravariant (or covariant) tensor over a vector space.
3) The ring of exterior differential polynomials in a differentiable manifold.

We define the *bar operation* in A to be an endomorphism of the additive group such that for $x \in A^k$,

$$\overline{x} = (-1)^k x. \tag{29}$$

Definition 6.3.2. An endomorphism ϑ of the additive group of a graded ring A is called a derivation if

$$\vartheta^{\overline{\alpha}} = \overline{\vartheta^{\alpha}}, \vartheta(\alpha\beta) = \vartheta(\alpha)\beta + \alpha\vartheta(\beta). \tag{30}$$

It is called an anti-derivation if

$$\vartheta^{\overline{\alpha}} = -\overline{\vartheta^{\alpha}}, \vartheta(\alpha\beta) = \vartheta(\alpha)\beta + \overline{\alpha}\vartheta(\beta). \tag{31}$$

The following theorem is easily verified:

Theorem 6.3.3. *Let A be a graded ring.*

1) *If ϑ is an anti-derivation, then $\vartheta\vartheta$ is a derivation.*
2) *If ϑ_1, ϑ_2 are anti-derivations, then $\vartheta_1\vartheta_2 + \vartheta_2\vartheta_1$ is a derivation.*
3) *Define*

$$[\vartheta_1, \vartheta_2] = \vartheta_1\vartheta_2 - \vartheta_2\vartheta_1. \tag{32}$$

If ϑ_1, ϑ_2 are derivations, then $[\vartheta_1, \vartheta_2]$ is a derivation. If ϑ_1 is derivation and ϑ_2 an anti-derivation, then $[\vartheta_1, \vartheta_2]$ is an anti-derivation.

We now give some examples of anti-derivations. An important example of derivation will be found in the next section.

Example 6.3.4. Exterior differentiation in the ring of exterior polynomials in a differentiable manifold is an anti-derivation.

Example 6.3.5. Let V be a vector space of dimension n.

We take a fixed element x' of the dual space V^* of V and consider the mapping in the exterior algebra $\wedge(V)$ defined by

$$\vartheta\alpha \equiv i(x')\alpha \equiv \alpha \lfloor x', \qquad \alpha \in \wedge(V).$$

We shall prove that this mapping is an anti-derivation. It is clearly an endomorphism in the additive group. We need only verify the relations (31) for the case where α, β are homogeneous. If α is homogeneous of degree k, we have, for any $z' \in \wedge^{k-1}(V^*)$,

$$\langle i(x')\overline{\alpha}, z'\rangle = \langle \overline{\alpha}, x' \wedge z'\rangle = (-1)^k \langle \alpha, x' \wedge z'\rangle,$$
$$\langle \overline{i(x')\alpha}, z'\rangle = (-1)^{k-1}\langle i(x')\alpha, z'\rangle = (-1)^{k-1}\langle \alpha, x' \wedge z'\rangle.$$

Hence

$$\overline{i(x')\alpha} = -i(x')\overline{\alpha},$$

which proves the first equation of (31).

To prove the second equation of (31) it is sufficient to consider the case that α, β are homogeneous of degrees k and ℓ respectively. If $k = \ell = 1$, we have, for any $z' \in \wedge(V^*)$,

$$\begin{aligned}\langle i(x')(\alpha \wedge \beta), z'\rangle &= \langle \alpha \wedge \beta, x' \wedge z'\rangle \\ &= \langle \alpha, x'\rangle\langle \beta, z'\rangle - \langle \alpha, z'\rangle\langle \beta, x'\rangle,\end{aligned}$$
$$\langle (i(x')\alpha) \wedge \beta, z'\rangle = \langle \alpha, x'\rangle\langle \beta, z'\rangle,$$
$$\langle \overline{\alpha} \wedge (i(x')\beta), z'\rangle = -\langle \beta, x'\rangle\langle \alpha, z'\rangle,$$

so that the relation in question is true. To established the formula in the general case we apply double induction on k, ℓ. Suppose the formula be true for k and $\ell - 1$. It suffices to consider the case that β is of the form

$$\beta = \beta_1 \wedge \beta_2,$$

where β_1, β_2 are homogeneous of degrees 1 and $\ell - 1$ respectively. For such α, β we have, on applying the induction hypothesis,

$$\begin{aligned}i(x')(\alpha \wedge \beta) &= i(x')(\alpha \wedge \beta_1 \wedge \beta_2) \\ &= (i(x')(\alpha \wedge \beta_1)) \wedge \beta_2 + \overline{\alpha \wedge \beta_1} \wedge (i(x')\beta_2) \\ &= \{i(x')\alpha \wedge \beta_1 + \overline{\alpha} \wedge i(x')\beta_1\} \wedge \beta_2 + \overline{\alpha} \wedge \overline{\beta_1} \wedge i(x')\beta_2,\end{aligned}$$

$$i(x')\alpha \wedge \beta + \overline{\alpha} \wedge i(x')\beta = i(x')\alpha \wedge \beta_1 \wedge \beta_2 + \overline{\alpha} \wedge \{i(x')\beta_1 \wedge \beta_2 + \overline{\beta}_1 \wedge i(x')\beta_2\}.$$

It follows that the formula is true for k, ℓ. Similarly, we prove the induction on k. Thus the second relation in (31) is valid in general. We have therefore proved that the mapping $\alpha \longrightarrow i(x')\alpha$ is an anti-derivation.

We state here the following useful property of derivations, which follows immediately from the definition:

Theorem 6.3.6. *If a graded ring A is generated by its homogeneous elements of degrees O and* 1, *any two derivations or anti-derivations which agree on these elements are identical.*

6.4 Infinitesimal transformation

We recall that an infinitesimal transformation X of class k is by definition a contravariant vector field of class k. This section will be devoted to the notions which arise when such an X is given. The discussions being local, we suppose that $X \neq 0$ in the neighbourhood under consideration.

Let $p \in M$ and let U be a coordinate neighbourhood of p, in which an infinitesimal transformation $X \neq 0$ is given. This defines a differential system of dimension 1 in U. By the Lemma in §6.2, such a differential system is always completely integrable. There exists therefore at p a local coordinate system $v^1, \ldots, v^n$ such that $v^2 =$ const., ..., $v^n =$ const. are integral manifolds. Moreover, the coordinate v^1 can be so chosen that the corresponding parametrized curves with v^1 as parameter have at every point q the vector $X(q)$ as tangent vector. Writing t for v^1, we see that the equations

$$u^i = u^i(v^2, \ldots, v^n, t), \ i = 1, \ldots, n, \tag{33}$$

which define the coordinate transformations at p, are the equations of the integral curves (i.e., integral manifolds of dimension 1), when the v's are regarded as constants and t as the parameter. These integral curves are solutions of the system of ordinary differential equations

$$\frac{du^i}{dt} = X^i, \tag{34}$$

where X^i are the components of X relative to the natural base in U. The parameter t here is defined up to an additive constant. The equations (33) may be pictured geometrically as defining a "flow" in the neighbourhood of p.

We shall choose t so that $t = 0$ at p. There are a neighbourhood W of p on U and an $\varepsilon > 0$ such that for $|t| < \varepsilon$, the coordinates $(v^1 + t, v^2, \ldots, v^n)$ are the v-coordinates of a point q' in U, when $(v^1, \ldots, v^n)$ are the v-coordinates of a point $q \in W$. Let φ_t be the mapping defined by $q' = \varphi_t(q)$. As usual this induces linear maps on the tangent spaces and their dual spaces:

$$\begin{aligned} (\varphi_t)^* &: V(q) \longrightarrow V(q'), \ q \in W, \\ \varphi_t^* &: V^*(q') \longrightarrow V^*(q), \ q \in W. \end{aligned} \tag{35}$$

Now suppose a covariant tensor field T^* of type $(0,s)$ and class k be given in U. Then $(\varphi_t^*)_s T^*(q')$, where

$$(\varphi_t^*)_s = \varphi_t^* \otimes \cdots \otimes \varphi_t^*, \tag{36}$$

s factors is a tensor of the same type at q. We define a tensor field $\vartheta(X)T^*$ in W by

$$(\vartheta(X)T^*)(q) = \lim_{t \to 0} \frac{(\varphi_t^*)_s T^*(q') - T^*(q)}{t}. \tag{37}$$

It is of type $(0,s)$ and class min $(k-1, \ell)$, where ℓ is the class of X. In the ring of covariant tensor fields $\vartheta(X)$ is a derivation. For, if T_1^* and T_2^* are two covariant tensor fields, we clearly have

$$\vartheta(X)(T_1^* \otimes T_2^*) = \vartheta(X)T_1^* \otimes T_2^* + T_1^* \otimes \vartheta(X)T_2^*. \tag{38}$$

Similarly, we define the operation $\vartheta(X)$ in the ring of exterior polynomials in U and in the ring of contravariant tensor fields in U. In each case it is a derivation. It follows easily from definition that if Y is a vector field and w a linear differential form, we have

$$\vartheta(X)\langle Y, \omega \rangle = \langle \vartheta(X)Y, \omega \rangle + \langle Y, \vartheta(X)\omega \rangle. \tag{39}$$

These operations $\vartheta(X)$ can be extended to the ring of mixed tensor fields of type (r,s). In fact, such a tensor field can be written as $T_0 \otimes T^*$, where T is a contravariant tensor field of type $(r,0)$ and T^* is a covariant tensor field of type $(0,s)$. We therefore define

$$\vartheta(X)(T \otimes T^*) = \vartheta(X)T \otimes T^* + T \otimes \vartheta(X)T^*, \tag{40}$$

so that $\vartheta(X)$ again maps a tensor field to one of the same type.

In view of (38) it is sufficient to study the operation $\vartheta(X)$ in the following simple cases:

1) scalar field;
2) linear differential form ω or, more particularly, df;
3) infinitesimal transformation Y.

We proceed to derive their simplest properties.

Let f be a scalar function in U, i.e., a tensor field of type $(0,0)$. We assert that the action of $\vartheta(X)$ is given by the formulas

$$\vartheta(X)f = Xf, \tag{41}$$

$$\vartheta(X)df = dXf. \tag{42}$$

To prove these we observe that both sides of (41) and (42) are independent of the choice of the local coordinate system, so that it suffices to verify them relative to the coordinates v^i chosen above. In the coordinates v^i the vector X has the components $1, 0, \ldots, 0$, and we have

$$\begin{aligned}\vartheta(X)f &= \lim_{t\to 0}\frac{f(v^1+t,v^2,\ldots,v^n)-f(v^1,\ldots,v^n)}{t}\\ &= \frac{\partial f}{\partial v^1}\\ &= Xf\\ \vartheta(X)df &= \lim_{t\to 0}\frac{df(v^1+t,v^2,\ldots,v^n)-df(v^1,\ldots,v^n)}{t}\\ &= d\frac{\partial f}{\partial v^1}\\ &= dXf\end{aligned}$$

which proves (41) and (42).

For a given infinitesimal transformation Y, $\vartheta(X)Y$ is an infinitesimal transformation. We shall introduce the notation

$$\vartheta(X)Y = [X,Y] \tag{43}$$

and call it the *Poisson bracket* of X, Y. To a given scalar function f we have, by using (39), (41), (42),

$$\begin{aligned}[X,Y]f &= \langle [X,Y], df\rangle\\ &= \langle \vartheta(X)Y, df\rangle\\ &= \vartheta(X)\langle Y, df\rangle - \langle Y, \vartheta(X)df\rangle\\ &= \vartheta(X)Yf - \langle Y, dXf\rangle\\ &= XYf - YXf.\end{aligned}$$

Thus we get the following formula, which is usually taken as a definition of the Poisson bracket:

$$[X,Y]f = XYf - YXf. \tag{44}$$

From (44) we immediately get the following fundamental relations for the Poisson brackets:

$$[X,Y] = -[Y,X], \tag{45}$$

$$[X,[Y,Z]] + [Y,[Z,X]] + [Z[X,Y]] = 0. \tag{46}$$

Equation (46) is known as *Jacobi's identity.*

It may be remarked here that equations (41), (42), (44) can be taken as definitions of the operator $\vartheta(X)$. From these we can extend $\vartheta(X)$ to general tensor fields by assuming that it is a derivation. Adopting these formulas as definition, we see that it is not necessary to assume $X \neq 0$. This assumption will henceforth be dropped.

Regarding the effect of the operator $\vartheta(X)$ on exterior polynomials, we have the following formula:

$$\vartheta(X) = i(X)d + di(X). \tag{47}$$

To prove this we notice that both sides are derivations (Theorem 6.3.3) and that the ring of exterior polynomials is (locally) generated by the exterior differential forms of degree 0 and 1. By Theorem 6.3.6 it suffices to prove (47) for the forms of degrees 0 and 1. We can moreover assume the form of degree 1 to be df. Then we have

$$(i(X)d + di(X))f = i(X)df = \langle X, df\rangle = Xf = \vartheta(X)f,$$
$$(i(X)d + di(X))df = di(X)df = dXf = \vartheta(X)df,$$

which proves (47). An immediate consequence of (47) is the "commutativity" of $\vartheta(X)$ and d:

$$d\vartheta(X) = \vartheta(X)d. \tag{48}$$

The Poisson bracket operation is intimately related to exterior differentiation. In most cases one can be used in place of the other. Their relationship can be expressed by the formula:

$$\vartheta(X)\langle Y,\omega\rangle - \vartheta(Y)\langle X,\omega\rangle \;=\; \langle X\wedge Y, d\omega\rangle + \langle [X,Y],\omega\rangle. \tag{49}$$

where X, Y are infinitesimal transformations and ω is a linear differential form. To prove (49) we observe that it suffices to take $\omega = gdf$, where f, g and scalar functions. The left-hand side of (49) is then equal to

$$\begin{aligned}&2\langle [X,Y],\omega\rangle + \langle Y, \vartheta(X)\omega\rangle - \langle X, \vartheta(Y)\omega\rangle\\ =&2g\langle [X,Y], df\rangle + \langle Y, Xgdf + gdXf\rangle - \langle X, Ygdf + gdYf\rangle\\ =&g[X,Y]f + XgYf - XfYg + \langle [X,Y], gdf\rangle + \langle X\wedge Y, dg\wedge df\rangle,\end{aligned}$$

which is equal to the right-hand side of (49).

To visualize this relationship even more clearly we take locally the dual bases X_i, ω^k, so that

$$\langle X_i, \omega^k\rangle = \delta_i^k \tag{50}$$

(All indices in this paragraph run from 1 to n). We write

$$d\omega^i = \frac{1}{2}\sum_{1,m} \alpha^i_{1m}\omega^1\wedge\omega^m, \tag{51}$$

$$[X_j, X_k] = -\sum_i b^i_{jk}X_i, \tag{52}$$

where b^i_{jk}, c^i_{jk} are scalar functions in U satisfying

$$b^i_{jk} + b^i_{kj} = 0, \tag{53}$$

$$c^i_{jk} + c^i_{kj} = 0. \tag{54}$$

Using (49) we find

$$c^{i}_{jk} = b^{i}_{jk}. \tag{55}$$

Thus the coefficients in (51) and (52) are identical.

These considerations allow us to express the conditions in Theorem 6.2.4 in a different form. Let the bases be so chosen that the linear space $F(p)$ defined by the differential system F at p is spanned by $X_1, \ldots, X_r$. Then using the notations of the theorem, we can take

$$\Omega = \omega^{r+1} \wedge \cdots \wedge \omega^{n}.$$

The condition of complete integrability can, by (23), be expressed as

$$C^{\wedge}_{\alpha\beta} = 0, \qquad \alpha, \beta = 1, \ldots, r, \wedge = r+1, \ldots, n \tag{56}$$

This is equivalent to saying that $[X_\alpha, X_\beta]$ is a linear combination of $X_1, \ldots, X_r$ only. In other words, the *system F is completely integrable if and only if, for any two infinitesimal transformations X, Y such that $X(p)$, $Y(p) \in F(p)$ for all $p \in U$, the Poisson bracket $[X, Y](p) \in F(p)$.*

6.5 Integration of differential forms

In this section we shall develop the theory of integration of exterior differential forms on a differential manifold. For this purpose some preliminary concepts and lemmas will be needed.

Definition 6.5.1. A differentiable manifold M of dimension n is called *orientable* if there is a continuous exterior differential form of degree n which is nowhere zero. Two such forms define the same orientation if they differ from each other by a factor which is constantly positive.

It follows that an orientable manifold has exactly two orientations. For if ω, ω' are forms of degree n defining two orientations of M, the scalar function f defined by $\omega' = f\omega$ is either constantly positive or constantly negative. It defines the same orientation as ω or $-\omega$, according as the former or the latter is the case. The manifold M is called *oriented* if such a form $\omega \neq 0$ is given. This form defines then an orientation in each tangent space of M. A form of degree n can then be written as $f(p)\omega$; it will be said to be > 0, < 0, or $= 0$, at p, according as $f(p) > 0$, < 0, or $= 0$. This property depends only on the orientation of M, and not on the choice of ω to define the orientation.

Definition 6.5.2. The carrier of a scalar function f in M is the closure of the set of points of M at which f is not equal to zero. Its complement is therefore the maximal open set in which $f = 0$. In general, the carrier of a form is the closure of the set of points at which it is different from zero.

Definition 6.5.3. An open covering of M is called *locally finite*, if any compact set of M meets only a finite number of its members.

Theorem 6.5.4. *Let K be a family of open sets of M which form a base. There is a locally finite open covering of M whose members are sets of K.*

We shall first construct covering of M with special properties. For convenience we shall use the term "countable" to mean "countable or finite".

Since M is separable, there is a countable covering $\{P_i\}$ of M by open sets P_i with compact closures. Put

$$Q_j = \bigcup_{1 \le i \le j} \overline{P}_i.$$

Then $\{Q_j\}$ form a countable covering of M by compact sets Q_j such that $Q_j \subset Q_{j+1}$. We now construct compact sets R_j such that

$$Q_j \subset R_j,\ R_j = \text{Int. of } R_{j+1}.$$

This is done by induction. In fact, suppose $R_1, \ldots, R_j$ be constructed. Since $R_j \cup Q_{j+1}$ is compact, it has a finite covering by open sets with compact closures. We put R_{j+1} to be the union of these closures.

Let $S_j = \text{Int. of } R_j$, and let

$$T_j = R_j \cap (M - S_{j-1}),$$

where we agree that sets with negative indices are void. Since $R_{j-1} \subset S_{j-1}$ and $T_j \subset M - S_{j-1}$, we have

$$T_j \cap R_{j-2} = \Phi.$$

For $p \in T_j$ there is a set of K containing p, contained in S_{j+1}, and not meeting R_{j-2}. These sets, for all $p \in T_j$, form a covering of T_j. On the other hand, T_j is a closed subset of a compact set R_j, and is hence compact. It follows that the above covering has a finite sub-covering, which we call K_j. We denote by K' the family of the sets of K_j for all j.

The sets of K' form a covering of M. In fact, if $p \in M$, there is a $k > 0$ such that $p \in R_k$, $p \notin R_{k-1}$. Hence p belongs to T_k, and is covered by a set of K'. Moreover, if $j \ge k+2$, R_k meets no set of K_j. Since every compact set of M is contained in a certain P_k, it follows that the covering K' is locally finite.

Theorem 6.5.5. *(Partition of unity). Let $\{U_i\}$ be an open covering of a manifold M of class $k \ge 1$. There are functions g_α satisfying the conditions:*

1) *Each g_α is of class k and satisfies the inequalities $0 \le g_\alpha \le 1$. Its carrier is compact and contained in one of U_i.*
2) *Every point of M has a neighbourhood which meets only a finite number of the carriers of g_α.*
3) $\sum g_\alpha = 1$.

We need the following well-known lemma, which can be found for instance in Steenrod, *Fibre bundles*, p. 26:

Let V be an open set in M with compact closure $\overline{V}$, which is contained in a coordinate neighbourhood U. Then there exists a real-valued function h defined in M of class k such that $0 \le h(p) \le 1$ for all $p \in M$, $h(p) = 1$ for $p \in \overline{V}$, and $h(p) = 0$ for $p \in M - U$.

We now consider open sets W which have the following properties:

1) The closure $\overline{W}$ is compact and contained in a certain U_i.
2) There is a real-valued function w of class k in M, satisfying the inequalities $0 \le w \le 1$, such that W is the set of all the points at which $w \ne 0$.

By the above lemma it follows that the family of all such sets W constitutes a base of M. By Theorem 6.5.4 we see that there is a locally finite countable open covering of M formed by the sets W. Denote such a covering by $\{W_\alpha\}$, and the function corresponding to W_α by w_α. Since the covering is locally finite, every point $p \in M$ has a neighbourhood which meets only a finite number of W_α, i.e., in which only a finite number of the functions w_α takes non-zero values. The sum $\sum_\alpha w_\alpha(p)$ is therefore a finite sum and is different from zero. We define

$$g_\alpha(p) = \frac{w_\alpha(p)}{\sum_\alpha w_\alpha(p)}.$$

It is easy to verify that these functions satisfy the conditions of the theorem.

Theorem 6.5.6. *Let M be an oriented manifold of dimension n. There is one, and only one, functional which assigns to a continuous differential form Φ of degree n with compact carrier a real number, called its integral and to be denoted by $\int \Phi$, such that the following condition are satisfied:*

$$1) \qquad \int (\Phi_1 + \Phi_2) = \int \Phi_1 + \int \Phi_2.$$

If the carrier of Φ is contained in a coordinate neighbourhood U with the coordinates $u^1, \dots, u^n$ such that $du^1 \wedge \cdots \wedge du^n > 0$, and if, in terms of these coordinates,

$$\Phi = \varphi(u^1, \dots, u^n) du^1 \wedge \cdots \wedge du^n, \tag{57}$$

then

$$\int \Phi = \int_U \varphi(u^1, \dots, u^n) du^1 \dots du^n \tag{58}$$

where the right-hand side is a Riemann integral.

Suppose the form Φ be given, with compact carrier S. In order to define its integral we take an open covering $\{U_i\}$ of M by coordinate neighbourhood and a partition of unity g_α according to Theorem 6.5.5. Every point $p \in S$ has a neighbourhood V_p which meets only a finite number of the carriers of g_α. These V_p for all $p \in S$ form a covering of S. Since S its compact, it has a finite sub-covering. It follows that there is only a finite number of $g_\alpha \Phi$ which are not identically zero. We define

$$\int \Phi = \sum_{\alpha} \int g_{\alpha} \Phi, \tag{59}$$

where the right-hand side is a finite sum. The differential from in each summand having a carrier lying in a coordinate neighbourhood U_i, we can evaluate it according to (58).

In order that this definition of integral makes sense we have to prove that it is independent of the various choices made. These are:

1) Choice of the neighbourhood U_i which contains the carrier of $g_{\alpha}\Phi$;
2) Choice of the covering U_i and the corresponding partition of unity g_{α}.

Suppose the carrier of $g_{\alpha}\Phi$ be in two coordinate neighbourhood U_i, U_j, with the local coordinates $u^1, \ldots, u^n$, and $v^1, \ldots, v^n$ respectively. We can take aon open set W containing the carrier in which both coordinate systems are valid. In W we have

$$\begin{aligned} g_{\alpha}\Phi &= \varphi(u^1, \ldots, u^n) du^1 \wedge \cdots \wedge du^n \\ &= \psi(v^1, \ldots, v^n) dv^1 \wedge \cdots \wedge dv^n, \end{aligned}$$

where

$$\psi(v^1, \ldots, v^n) = \varphi(u^1(v^1, \ldots, v^n), \ldots, u^n(v^1, \ldots, v^n)) \wedge \frac{\partial(u^1, \ldots, u^n)}{\partial(v^1, \ldots, v^n)},$$

the functional determinant being constantly positive. The equation

$$\int_W \varphi du^1 \wedge \cdots \wedge du^n = \int_W \varphi \frac{\partial(u^1, \ldots, u^n)}{\partial(v^1, \ldots, v^n)} dv^1 \wedge \cdots \wedge dv^n$$

is then exactly the formula for the transformation of multiple integrals. Its validity proves the independence of the choice of U_i in the evaluation of the summands in (59).

Next consider a second covering $\{V_j\}$ of M by coordinate neighbourhoods, and let g'_{α} be a corresponding partition of unity. Then $\{U_i \cap V_j\}$ will be a covering of M, with the functions $g_{\alpha} g'_{\beta}$ as corresponding partition of unity. We find easily

$$\sum_{\alpha} \int g_{\alpha}\Phi = \sum_{\alpha,\beta} \int g_{\alpha} g'_{\beta} \Phi,$$

$$\sum_{\beta} \int g'_{\beta}\Phi = \sum_{\alpha,\beta} \int g_{\alpha} g'_{\beta} \Phi.$$

This proves the independence of the integral on the choice of the covering $\{U_i\}$ and the functions g_{α}.

It is clear that the integral so defined satisfies the conditions 1) and 2). Its uniqueness is also immediate.

6.6 Formula of Stokes

The so-called Stokes formula deals with a relationship between an integral over a domain and one over its boundary. We shall first make precise these notions.

Definition 6.6.1. Let M be a differentiable manifold of dimension n. A domain D with regular boundary is a point set of M whose points have one of the properties:

1) It has a neighbourhood belonging entirely to D;
2) There is a coordinate neighbourhood U about point question with the coordinates $u^1,\dots,u^n$ such that $U\cap D$ is the set of all points $q\in U$ with $u^n(q)\geq u^n(p)$.

Clearly these two properties are exclusive of each other. Points with property 1) are called interior points of D and points with property 2) are called its boundary points. The set of all boundary points of D is called its boundary and will be denoted by B.

Theorem 6.6.2. *The boundary of a domain with regular boundary is a closed submanifold regularly imbedded. If M is orientable, so is the boundary B.*

The set B is closed, since $B=\overline{D}-D$. It is by definition a closed submanifold of M and is regularly imbedded; on account of Theorem 5.4.8.

Suppose M be oriented. To $p\in B$ we choose the coordinate system $u^1,\dots,u^n$ at p in condition 2) of Definition 6.6.1 such that

$$du^1\wedge\cdots\wedge du^n>0$$

Since $u^1,\dots,u^{n-1}$ form a coordinate system on B at p, we orient B by requiring that $(-1)^n du^1\wedge\cdots\wedge du^{n-1}>0$ at p. This definition is independent of the coordinate system u^i. In fact, let $v^1,\dots,v^n$ be a second coordinate system at p with $dv^1\wedge\cdots\wedge dv^n>0$, and satisfying condition 2) of Def 6.6.1. Since B is defined by $u^n(q)=u^n(p)$ and by $v^n(q)=v^n(p)$, where q is a point in a neighbourhood of p, we see that in the equations of transformation of coordinates v^n is a function of u^n only. Moreover, at p, $\frac{\partial v^n}{\partial u^n}\geq 0$. It follows that

$$\frac{\partial(v^1,\dots,v^n)}{\partial(u^1,\dots,u^n)}=\frac{\partial(v^1,\dots,v^{n-1})}{\partial(u^1,\dots,u^{n-1})}\cdot\frac{\partial v^n}{\partial u^n}>0$$

at p. This proves that $(-1)^n dv^1\wedge\cdots\wedge dv^{n-1}>0$ on B at p.

We have therefore defined at every point of the $(n-1)$-dimensional manifold B the sign, of a non-zero form of degree $n-1$. The bundle $\wedge^{*n-1}(B)$ of $(n-1)$-forms over B has fibers which are homeomorphic to the Euclidean line E'. If we omit the form 0, the fiber will be $E'-0$, where 0 is the origin on E', and is therefore not connected. By defining the sign of a non-zero form, we take a choice at every point of B, of the two components of the $E'-0$ This is moreover done continuously As a consequence the bundle of positive $(n-1)$-forms over B has fibers which are homeomorphic to the Euclidean line. In this bundle a cross-section can be defined, which gives a positive $(n-1)$-form over B. This B is oriented in the sense se of Definition 6.5.1. This completes the proof of Theorem 6.6.2.

We consider a domain D with regular boundary B, and suppose that it is compact. Define the characteristic function $h(p)$, $p \in M$, by

$$h(p) = 1, \quad p \in D, \quad h(p) = 0, \quad p \in M - D. \tag{60}$$

To an n-form Φ in M we define its integral over D by the formula

$$\int_D \Phi = \int h\Phi. \tag{61}$$

The differential form in the right-hand side is not continuous. But its integral can be defined, as a partition of unity will reduce the integral to the Euclidean case in which such an integral is defined.

Theorem 6.6.3 (Stokes). *Let ω be a form of degree $n-1$ and class ≥ 1 in M. Then we have*

$$\int_D d\omega = \int_B \omega. \tag{62}$$

Strictly speaking the integral on the right-hand side is

$$\int_D \omega = \int_B i^* \omega, \tag{63}$$

where $i : B \to M$ is the identity mapping.

To prove formula (62) we take a countable open covering $\{U_i\}$ of M by coordinate neighbourhoods such that each member U_i either has no point in common with B or is a coordinate neighbourhood having property 2) of Definition 6.6.1. Let the functions g_α be a corresponding position of unity. Since B and D are both compact, each of them meets only a finite number of the carriers of g_α, and we have

$$\int_B \omega = \sum_\alpha \int_B g_\alpha \omega,$$

$$\int_D d\omega = \sum_\alpha \int_D d(g_\alpha \omega).$$

It therefore suffices to prove (62) for each of the summands of the above sums, that is, under the assumption that the carrier of ω lies in an U_i.

Let $u^1, \ldots, u^n$ be the local coordinates in U_i such that $du^1 \wedge \cdots \wedge du^n > 0$, and let

$$\omega = \sum_{j=1}^{n} (-1)^{j-1} a_j du^1 \wedge \cdots \wedge du^{j-1} \wedge du^{j+1} \wedge \cdots \wedge du^n$$

in these coordinates, where a_j are functions of class ≥ 1. Then we have

$$d\omega = \left(\sum_{j=1}^{n} \frac{\partial a_j}{\partial u^j} \right)_{du^1 \wedge \cdots \wedge du^n} .$$

Suppose that u_i has no point in common with B. Then the right-hand side of (62) is zero. The set u_i either belongs to $M - D$ or to the interior of D. In the former case we have $h = 0$, so that (62) is true. In the latter case we have $h = 1$, and the integral in the left-hand side of (62) is equal to the integral

$$\int_C d\omega = \int_C \left(\sum_{j=1}^{n} \frac{\partial a_j}{\partial u^j} \right) du^1, \ldots, du^n,$$

where C is a cube in the space of the coordinates u^j which contains the carrier of ω in its interior. By choosing m sufficiently large, we can suppose C to be defined by the inequalities $|u^j| \leq m$. Applying the theorem on iterated integrals, we get

$$\begin{aligned} \int_C \frac{\partial a_j}{\partial u^j} du^1, \ldots, du^j, \ldots, du^n = \int_{\pm} \Big\{ & a_j(u^1, \ldots, u^{j-1}, m, u^{j+1}, \ldots, u^n) \\ & - a_j(u^1, \ldots, u^{j-1}, -m, u^{j+1}, \ldots, u^n) \Big\} \\ & du^1 \cdots du^{j-1} du^{j+1} \cdots du^n \\ = & 0. \end{aligned}$$

Hence in this case the left-hand side of (62) is again zero.

Next suppose U_i be a coordinate neighbourhood having property 2) of Definition 6.6.1. We assume B to be contained in the subset defined by $u^n = 0$. Then $h(p) = 1$ when $u^n(p) \geq 0$. In the space of the coordinates u^j we take a cube C defined by

$$|u^k| \leq m, \quad k = 1, \ldots, n-1, \quad m \geq u^n \geq 0,$$

such that the carrier of ω is contained in the union of its in interior and the side $u^n = 0$. As in the above, we have

$$\int_C \frac{\partial a_k}{\partial u^k} du^1, \ldots, du^n = 0, \quad k = 1, \ldots, n-1.$$

Similarly we have

$$\int_C \frac{a_n}{\partial u^n} du^1, \ldots, du^n = (-1)^{n-1} \int_B a_n(u^1, \ldots, u^{n-1}, 0) du^1, \ldots, du^{n-1}.$$

But the right-hand side of the last equation is equal to

$$\int_B \omega.$$

This proves (62) for the case under consideration.

7

Affine Connections

An affine connection on a differentiable manifold is a structure which enables us to "differentiate" tensor fields. For simplicity we shall assume our manifold to be of class ∞, although most results will be valid under weaker hypotheses. We shall also agree that all small Latin indices run from 1 to n, unless otherwise stated.

7.1 Definition of an affine connection: covariant differential

Let f be a differentiable function defined over an open subset U of a differentiable manifold M (of dimension n). Then df is a covariant vector field and we have, relative to a local coordinate system u^i,

$$df = \sum_i \frac{\partial f}{\partial u^i} du^i. \tag{1}$$

In other words, the partial derivatives of f with respect to u^i are the components of a covariant vector field relative to the corresponding natural base. This process of getting new tensor fields by taking the partial derivatives of the components of a tensor field with respect to a local coordinate system cannot be extended to tensor fields of higher orders. In fact, consider a contravariant vector field X whose components in two local coordinate systems u^i and u^{*i} are X^i and X^{*i} respectively, so that we have

$$X = \sum_i X^i \frac{\partial}{\partial u^i} = \sum_i X^{*i} \frac{\partial}{\partial u^{*i}}, \tag{2}$$

or

$$X^{*i} = \sum_j \frac{\partial u^{*i}}{\partial u^j} X^j. \tag{3}$$

Partial differentiation gives

$$\frac{\partial X^{*i}}{\partial u^{*k}} = \sum_{j,\ell} \frac{\partial u^{*i}}{\partial u^j} \frac{\partial u^\ell}{\partial u^{*k}} \frac{\partial X^j}{\partial u^\ell} + \sum_{j,\ell} \frac{\partial^2 u^{*i}}{\partial u^j \partial u^\ell} \frac{\partial u^\ell}{\partial u^{*k}} X^j,$$

which shows that $\partial X^i/\partial u^k$ are not the components of a tensor field of type $(1,1)$.

To remedy this situation we introduce in M an affine connection. Historically the notion was guided by experience from Riemannian Geometry. With implications which will soon be clear we define it as follows:

Definition 7.1.1. An affine connection is defined by a set of linear differential forms ω^i_k in each coordinate neighbourhood, such that, in the intersection of two coordinate neighbourhoods, we have

$$dp^i_j = \sum_k \omega^{*i}_k p^k_j = \sum_k p^i_k \omega^k_j \tag{4}$$

where

$$p^i_j = \frac{\partial u^{*i}}{\partial u^j}, \tag{5}$$

the u^i's and the u^{*i}'s being the respective local coordinates.

We have to prove that the definition is consistent. By this we mean that, in an open subset in which three local coordinate systems u^i, u^{*i}, u^{**i} are valid, one of the relations (4) is a consequence of the two others. Suppose, in fact, that

$$d\left(\frac{\partial u^{*i}}{\partial u^j}\right) + \sum_k \omega^{*i}_k \frac{\partial u^{*k}}{\partial u^j} = \sum_k \frac{\partial u^{*i}}{\partial u^k} \omega^k_j,$$

$$d\left(\frac{\partial u^{**i}}{\partial u^{*j}}\right) + \sum_k \omega^{**i}_k \frac{\partial u^{**k}}{\partial u^{*j}} = \sum_k \frac{\partial u^{**i}}{\partial u^{*k}} \omega^{*k}_j,$$

we find

$$\begin{aligned}
& d\left(\frac{u^{**i}}{du^i}\right) + \sum_k \omega^{**i}_k \frac{\partial u^{**k}}{\partial u^j} \\
=& d\left(\sum_k \frac{\partial u^{**i}}{\partial u^{*k}} \frac{\partial u^{*k}}{\partial u^j}\right) + \sum_{k,\ell} \omega^{**i}_\ell \frac{\partial u^{**\ell}}{\partial u^{*k}} \frac{\partial u^{*k}}{\partial u^j} \\
=& \sum_k \frac{\partial u^{**i}}{\partial u^{*k}} \left\{ d\left(\frac{\partial u^{*k}}{\partial u^j}\right) + \sum_\ell \omega^{*k}_\ell \frac{\partial u^{*\ell}}{\partial u^j} \right\} \\
=& \sum_k \frac{\partial u^{**i}}{\partial u^k} \omega^k_j,
\end{aligned}$$

which proves the consistency of the definition.

Condition (4) can be put in an equivalent form, involving the quantities

$$q_k^i = \frac{\partial u^i}{\partial u^{*k}}. \tag{6}$$

In fact, the matrices

$$(p_i^j), \qquad (q_i^j)$$

are inverses to each other, so that we have

$$\sum_j p_i^j q_j^k = \sum_j q_i^j p_j^k = \delta_i^k. \tag{7}$$

Differentiating any one of these sets of equations, we get, after simplification

$$dq_j^i + \sum_k \omega_k^i q_j^k = \sum_k q_k^i \omega_j^{*k}. \tag{8}$$

These conditions are clearly equivalent to (4).

A differentiable manifold always has an affine connection. This can either be proved directly by means of some elementary extension theorems or by using the fact that M can always be given a Riemann metric, which in turn defines an affine connection, as will be established in the next chapter. In the present chapter we shall only be concerned with properties of M arising from a given affine connection. A manifold with an affine connection is sometimes called an *affinely connected manifold.* From now on we shall suppose on affine connection to be given in M.

By differentiating (3) we get, using (4),

$$\begin{aligned} dX^{*i} &= \sum_j p_j^i dX^j + \sum_j dp_j^i X^j \\ &= \sum_j p_j^i dX^j + \sum_j \left(\sum_k p_k^i \omega_j^k - \sum_k \omega_k^{*i} p_j^k \right) X^j, \end{aligned}$$

which can be written

$$dX^{*i} + \sum_k \omega_k^{*i} X^{*k} = \sum_j p_j^i \left(dX^j + \sum_k \omega_k^j X^k \right). \tag{9}$$

We define

$$DX^i = \sum_j X^i_{,j} du^j = dX^i + \sum_j \omega_j^i X^j. \tag{10}$$

Then $X^i_{,j}$ are the components of a tensor field of type $(1,1)$. Writing (9) in the form

$$DX^{*i} = \sum_j p_j^i DX^j, \tag{11}$$

we see that DX^i are linear differential forms behaving like a contravariant vector. They are said to define the *covariant differential of the vector field X*.

More generally, let, for instance, X_k^{ij} be the components of a tensor field of type $(2,1)$, relative to the local coordinates u^i. Under a change of local coordinates they are transformed as follows:

$$X_k^{*ij} = \sum_{\ell,m,r} p_\ell^i p_m^j q_k^r X_r^{\ell m}. \tag{12}$$

Making use of (4) and (8), we get, in the same way as above,

$$DX_k^{*ij} = \sum_{\ell,m,r} p_\ell^i p_m^j q_k^r DX_r^{\ell m}, \tag{13}$$

where

$$\begin{aligned} DX_k^{ij} &= \sum_\ell X_{k,\ell}^{ij} du^\ell \\ &= dX_k^{ij} + \sum_\ell \omega_\ell^i X_k^{\ell j} + \sum_\ell \omega_\ell^j X_k^{i\ell} - \sum_\ell \omega_k^\ell X_\ell^{ij}. \end{aligned} \tag{14}$$

It follows that the components $X_{k,\ell}^{ij}$ define a tensor field of type $(2,2)$. We shall say that $DX_k^{i,j}$ define the *covariant differential* of the tensor field X_k^{ij}. It is clear that these considerations extend to a tensor of any type. In particular, for a scalar function (tensor field of type $(0,0)$), we define its covariant differential to be the ordinary differential.

Under tensor multiplication and contraction of tensor fields we have, for instance,

$$D\left(\sum_{i,\ell} X_{\ell m}^{ijk} Y_i^\ell\right) = \sum_{i,\ell} DX_{\ell m}^{ijk} Y_i^\ell + \sum_{i,\ell} X_{\ell m}^{ijk} DY_i^\ell. \tag{15}$$

This formula is easily verified by computation.

If we introduce the quantities Γ_{jk}^i by the formulas

$$\omega_k^i = \sum_j \Gamma_{jk}^i du^j, \tag{16}$$

condition (4) can be written

$$\Gamma_{jk}^i = \sum_\ell \frac{\partial^2 u^{*\ell}}{\partial u^j \partial u^k} \frac{\partial u^i}{\partial u^{*\ell}} + \sum_{\ell,m,q} \Gamma_{mq}^{*\ell} \frac{\partial u^{*m}}{\partial u^j} \frac{\partial u^{*q}}{\partial u^k} \frac{\partial u^i}{\partial u^{*\ell}}. \tag{17}$$

These are the classical formulas of transformation of the "components" Γ_{jk}^i of an affine connection.

Definition 7.1.2. Let C be a parametrized curve in M, with the parameter t. The tangent vectors

$$X(t) = \sum_i X^i(t) \frac{\partial}{\partial u^i} \tag{18}$$

are said to be *parallel along* C if DX^i induces the form on C. Analytically the condition can be written

$$DX^i = 0, \tag{19}$$

$$\frac{dx^i}{dt} + \sum_{j,k} \Gamma^i_{jk} \frac{du^j}{dt} X^k = 0. \tag{20}$$

A parametrized curve is called an *auto-parallel curve*, or a *path*, if its tangent vectors are parallel along C.

Equations (20) are a system of ordinary differential equations of the first order. There exists, therefore one and only one solution $X(t)$ corresponding to any initial value $X(t_0)$. We shall interpret this geometrically by saying that the vector $X(t_0)$ is *displaced parallelly* along C.

By definition, the paths are integral curves of the system of differential equations

$$\frac{d^2u^i}{dt^2} + \sum_{j,k} \Gamma^i_{jk} \frac{du^j}{dt} \frac{du^k}{dt} = 0. \tag{21}$$

There is one, and only one, set of solutions $u^i(t)$, corresponding to given initial values $u^i(t_0)$, $\frac{du^i}{dt}(t_0)$. In other words, there is exactly one path through a point and tangent to a given vector.

The covariant differential of a tensor field leads to a new tensor field with one more covariant index. For instance, we have, from (14), the tensor field with the components $X^{ij}_{k,\ell}$ obtained from the tensor field X^{ij}_k. Applying this process again, we see that the components $X^{ij}_{k,\ell,m}$ defined by

$$DX^{ij}_{m,\ell} = \sum_m X^{ij}_{k,\ell,m} du^m \tag{22}$$

are those of a tensor field of type $(2,3)$. If will be of importance to find a formula for the difference

$$X^i_{jk,\ell,m} - X^i_{jk,m,\ell}.$$

Consider first a scalar function f. Relative to a local coordinate system u^i, we find

$$f_{,i} = \frac{\partial f}{\partial u^i},$$

$$f_{,i,j} = \frac{\partial^2 f}{\partial u^i \partial u^j} = \sum_k \Gamma^k_{ji} \frac{\partial f}{\partial u^k}.$$

It follows that

$$f_{,i,j} - f_{,j,i} = \sum_k (\Gamma^k_{ij} - \Gamma^k_{ji}) \frac{\partial f}{\partial u^k}. \tag{23}$$

Putting

$$T^k_{ij} = \Gamma^k_{ij} - \Gamma^k_{ji}, \tag{24}$$

we see that, for any scalar function f, the components

$$\sum_k T^k_{ij} \frac{\partial f}{\partial u^k}$$

are those of a tensor field of type $(0,2)$. This implies that T^k_{ij} are the components of a tensor field of type $(1,2)$. It depends only on the affine connection, and is called its *torsion tensor*. By (24) the torsion tensor is anti-symmetric in the covariant indices.

Next consider a contravariant vector field with the components X^i. We find, by definition

$$X^i_{,j} = \frac{\partial X^i}{\partial u^j} + \sum_j \Gamma^i_{j\ell} X^\ell,$$

$$X^i_{,j,k} = \sum_\ell \left(\frac{\partial \Gamma^i_{j\ell}}{\partial u^k} + \sum_m \Gamma^i_{km} \Gamma^m_{j\ell} \right) X^\ell - \sum_\ell \Gamma^\ell_{kj} X^i_{,\ell}$$
$$+ \text{ terms symmetric in } j,k.$$

It follows that

$$X^i_{,j,k} - X^i_{,k,j} = \sum_\ell T^\ell_{jk} X^i_{,\ell} + \sum_\ell R^i_{j\ell k} X^\ell, \tag{25}$$

where

$$R^i_{\ell jk} = \frac{\partial \Gamma^i_{k\ell}}{\partial u^j} - \frac{\partial \Gamma^i_{j\ell}}{\partial u^k} + \sum_m (\Gamma^i_{jm} \Gamma^m_{k\ell} - \Gamma^i_{km} \Gamma^m_{j\ell}). \tag{26}$$

The $R^i_{\ell jk}$ are the components of a tensor field of type $(1,3)$, anti-symmetric in its last two indices. It depends only on the affine connection, and is called its *curvature tensor*.

The derivation of similar formulas for general tensor fields does not give rise to new tensors of the affine connection. We shall give them in the next section, where the computation will be simplified after more praparations.

A trivial but important example of affine connection is given by the ordinary affine space A^n. As a differentiable manifold it is the n-dimensional vector space, with the usual differentiable structure. The manifold is, therefore, covered by one coordinate neighbourhood and the affine connection is defined by setting all the components Γ^i_{jk} equal to zero. From (24) and (26) both its torsion tensor and curvature tensor are zero. At any point $p \in A^n$ the linear map which sends the tangent vector $\frac{\partial}{\partial u^i}$ to the vector $(0,\ldots,0,1,0,\ldots,0)$ (1 in the ith place) identifies the tangent space $V(p)$ at P with A^n itself. Two tangent vectors, at the points p and q respectively, are parallel if and only if they are identified to the same vector of A^n. In this case, therefore, the parallelism of vectors is independent of the curve joining them; in general this is not so.

7.2 The principal bundle

Associated to a differentiable manifold there are two fiber bundles of particular importance; the tangent bundle of its tangent vectors and the so-called principal fiber bundle. When an affine connection is given in the manifold, this has implications for these bundles, which we shall now consider.

As usual, let M be a differentiable manifold of dimension n. We shall call a *frame* the object formed by a point $p \in M$ and n linearly independent tangent vectors at p. The principal bundle $\mathscr{B}$ is the space of all frames over M; its dimension is n^2+n. To a local coordinate system u^i in M there corresponds a system of local coordinates u^i, X_i^k in $\mathscr{B}$ defined by the condition that the n vectors of the frame are given by

$$\ell_i = \sum_k X_i^k \frac{\partial}{\partial u^k}. \tag{27}$$

Since these vectors are linearly independent, we have

$$\det(X_i^k) \neq 0, \tag{28}$$

and we can introduce the elements Y_i^k of the matrix inverse to (X_i^k) such that

$$\sum_j X_i^j Y_j^k = \sum_j Y_i^j X_j^k = \delta_i^k. \tag{29}$$

Suppose we restrict ourselves to a neighbourhood in which the u^i and a second local coordinate system u^{*i} are both valid. Denote the local coordinate system in $\mathscr{B}$ corresponding to u^{*i} by u^{*i}, X_i^{*k}, and introduce the Y_i^{*k}.

Then we have

$$\begin{aligned} X_i^{*k} &= \sum_j \frac{\partial u^{*k}}{\partial u^j} X_i^j, \\ Y_k^{*i} &= \sum_j Y_j^i \frac{\partial u^j}{\partial u^{*k}}, \end{aligned} \tag{30}$$

which implies in particular

$$\sum_j Y_j^{*i} du^{*j} = \sum_j Y_j^i du^j. \tag{31}$$

It follows that the differential form ϑ^i having as representatives the two sides of the above equation is independent of the choice of local coordinates and is defined in $\mathscr{B}$.

Suppose now an affine connection to be given in M. In the expression for DX^i in (10) we regard the X^i as independent variables and apply it to each of the vectors of our frame. Then

$$DX_i^j = dX_i^j + \sum_k \omega_k^k X_i^k \tag{32}$$

are linear differential forms in u^i, X_i^k. From (11) we get

$$DX_i^{*j} = \sum_k \frac{\partial u^{*j}}{\partial u^k} DX_i^k. \tag{33}$$

This, together with (30), gives

$$\sum_j Y_j^{*k} DX_i^{*j} = \sum_j Y_j^k DX_i^j. \tag{34}$$

It follows that the two members of this equation are representatives, in the coordinates u^i and u^{*i} respectively, of differential forms in $\mathscr{B}$. We shall denote them by ϑ_i^k. Notice that, whereas the ϑ^i are defined in $\mathscr{B}$ by means of the differential structure in M only, the ϑ_j^k are defined only when M has an affine connection. It is clear that these n^2+n linear differential forms ϑ^i, ϑ_i^k are linearly independent.

We shall find the formulas giving the exterior derivatives of these differential forms. Relative to the local coordinates u^i let us write

$$\vartheta^i = \sum_j Y_j^i du^j, \quad \vartheta_i^k = \sum_j Y_j^k \left(dX_i^j + \sum_\ell \omega_\ell^j X_i^\ell \right), \tag{35}$$

of

$$du^i = \sum_j X_j^i \vartheta^j, \qquad \sum_j X_j^k \vartheta_i^j = dX_i^k + \sum_\ell \omega_\ell^k X_i^\ell.$$

Applying exterior differentiation to these equations and simplifying, we get respectively

$$\sum_j X_j^i \left(d\vartheta^j - \sum_k \vartheta^k \wedge \vartheta_k^j \right) = \sum_j \omega_j^i \wedge du^j = \sum_{j,k} T_{jk}^i du^j \wedge du^k, \tag{36}$$

$$\sum_j X_j^k \left(d\vartheta_i^j - \sum_\ell \vartheta_i^\ell \wedge \vartheta_\ell^j \right) = \sum_j X_i^j \left(d\omega_j^k - \sum_\ell \omega_j^\ell \wedge \omega_\ell^k \right)$$

$$= \frac{1}{2} \sum_{j,\ell,m} X_i^j R_{j\ell m}^k du^\ell \wedge du^m,$$

or

$$d\vartheta^j - \sum_k \vartheta^k \wedge \vartheta_k^j = \frac{1}{2}\sum_{\ell m} P_{\ell m}^j \vartheta^\ell \wedge \vartheta^m, \tag{37}$$

$$d\vartheta_i^j - \sum_\ell \vartheta_i^\ell \wedge \vartheta_\ell^j = \frac{1}{2}\sum_{\ell,m} S_{i\ell m}^j \vartheta^\ell \wedge \vartheta^m,$$

where

$$\frac{1}{2}P_{\ell m}^j = \sum_{i,p,q} Y_i^j X_\ell^p X_m^q T_{pq}^i, \qquad S_{i\ell m}^j = \sum_{k,p,q,r} Y_k^j X_i^p X_\ell^q X_m^r R_{pqr}^k. \tag{38}$$

Equations (37) play an extremely important role in the theory of affine connections. We shall call them the *equations of structure*. Since both sides of (37) do not involve the local coordinates, they are valid in the whole space $\mathscr{B}$. The coefficients $P_{\ell m}^j$, $S_{i\ell m}^j$ are functions in $\mathscr{B}$, i.e., are functions of the frames.

In particular, these equations remain valid when we put

$$X_i^k = Y_i^k = \delta_i^k. \tag{39}$$

This is done relative to a local coordinate system u^i, and conditions (39) signify that we restrict ourselves to the natural frames. Under this restriction formulas (37) reduce to

$$-\sum_k du^k \wedge \omega_k^j = \sum_{\ell,m} T_{\ell m}^j du^\ell \wedge du^m, \tag{40}$$

$$d\omega_i^j - \sum_\ell \omega_i^\ell \wedge \omega_\ell^j = \frac{1}{2}\sum_{\ell,m} P_{i\ell m}^j du^\ell \wedge du^m.$$

It is sometimes convenient to introduce the exterior quadratic differential forms

$$\Theta^j = \frac{1}{2}\sum_{\ell,m} P_{\ell m}^j \vartheta^\ell \wedge \vartheta^m, \tag{41}$$

$$\Theta_i^j = \frac{1}{2}\sum_{\ell,m} s_{i\ell m}^j \vartheta^\ell \wedge \vartheta^m,$$

and write equations (37) as

$$d\vartheta^j - \sum_k \vartheta^k \wedge \vartheta_k^j = \Theta^j, \tag{42}$$

$$d\vartheta_i^j - \sum_k \vartheta_i^k \wedge \vartheta_k^j = \Theta_i^j.$$

Exterior differentiation of these equations gives

$$d\Theta^j + \sum_k \Theta^k \wedge \vartheta_k^j - \sum_k \vartheta^k \wedge \Theta_k^j = 0, \tag{43}$$

$$d\Theta_i^j + \sum_k \Theta_i^k \wedge \vartheta_k^j - \sum_k \vartheta_i^k \wedge \Theta_k^j = 0.$$

Equations (43) are called the *Bianchi identities.*

We have shown in the above that the existence of an affine connection in M gives rise to $n^2 + n$ linearly independent linear differential forms ϑ^i, ϑ_i^k in $\mathscr{B}$ whose exterior derivatives are given by simple formulas. The question naturally arises how a set of differential forms ϑ_i^k should behave in $\mathscr{B}$ in order to define an affine connection in M. This is answered by the following theorem.

Theorem 7.2.1. *In order that the linear differential forms ϑ_i^j in $\mathscr{B}$ define an affine connection in M it is necessary and sufficient that they, together with ϑ^i, satisfy the equations of structure* (37).

It remains only to prove the sufficiency. Since ϑ^i are given in $\mathscr{B}$ by the differentiable structure of M, the problem is to determine ϑ_i^j so that equations (37) are satisfied. We make use of a local coordinate system u^i in M, and the corresponding local coordinates u^i, X_i^k in $\mathscr{B}$. Using the expressions in (35) for ϑ^i, we find

$$\begin{aligned} d\vartheta^i &= \sum_k dY_k^i \wedge du^k = -\sum_{j,k,\ell} Y_j^i dX_\ell^j \wedge Y_k^\ell du^k \\ &= -\sum_{j,\ell} Y_j^i dX_\ell^j \wedge \vartheta^\ell, \end{aligned}$$

so that the first set of equations in (37) can be written

$$\sum_j \vartheta^j \wedge \left(\vartheta_j^i + \frac{1}{2} \sum_\ell P_{j\ell}^i \vartheta^\ell - \sum_\ell Y_\ell^i dX_j^\ell \right) = 0.$$

This implies that $\vartheta_j^i - \sum_\ell Y_\ell^i dX_j^\ell$ are linear combinations of ϑ^k, and, guided by the second set of formulas in (35), we shall put

$$\sum_j X_j^k \vartheta_i^k - dX_i^k = \sum_j \omega_j^k X_i^j, \tag{44}$$

where ω_j^k are linear combinations of du^i. The theorem will be established, if we prove that ω_j^k are linear differential forms in u^i only. Since ω_j^k can be written in the form

$$\omega_j^k = \sum_i \Gamma_{ij}^k du^i,$$

the problem is to prove that the Γ_{ij}^k are independent of X_ℓ^m.

For this purpose we take the exterior derivative of the equation (44), obtaining

$$\sum_j d\omega_j^k X_i^j - \sum_j \omega_j^k \wedge dX_i^j$$
$$= \sum_j dX_j^k \wedge \vartheta_i^j + \sum_j X_j^k d\vartheta_i^j.$$

Using (44) and (37), we can simplify this equation to

$$\sum_j X_i^j \left(d\omega_j^k - \sum_\ell \omega_j^\ell \wedge \omega_\ell^k \right)$$
$$= \frac{1}{2} \sum_{j,\ell,m} X_j S_{i\ell m}^j \vartheta^\ell \wedge \vartheta^m. \tag{45}$$

This implies that $d\omega_j^k$ is an exterior quadratic differential form in du^i.

On the other hand, we have

$$d\omega_j^k = \sum_{i,\ell} \frac{\partial \Gamma_{ij}^k}{\partial u^\ell} du^\ell \wedge du^i$$
$$+ \sum_{i,\ell,m} \frac{\partial \Gamma_{ij}^k}{\partial X_\ell^m} dX_\ell^m \lambda du^i.$$

It follows that

$$\frac{\partial \Gamma_{ij}^k}{\partial X_\ell^m} = 0,$$

so that Γ_{ij}^k are independent of X_ℓ^m. Thus Theorem 7.2.1 is proved.

We now study the notion of a tensor field from the point of view of the principal bundle. For definiteness consider a tensor field of type (2, 1). Its components in terms of two local coordinate systems are related by the equations (12). These, together with equations (30), give

$$\sum_{q,r,s} Y_q^{*i} Y_r^{*j} X_k^{*s} X_s^{*qr} = \sum_{q,r,s} Y_q^j Y_r^i X_k^s X_s^{qr}. \tag{46}$$

The common expression of this equation is, therefore, a function in $\mathscr{B}$, independent of the choice of local coordinates. There are altogether n^3 such functions, which we shall denote by T_k^{ij}. In general, for a tensor field of type (k, ℓ) there will correspond in $\mathscr{B}$ $n^{k+\ell}$ functions.

When does such a set of functions T_k^{ij} arise from a tensor field? A necessary and sufficient condition is that they have the expression in the right hand side of (46) in the local coordinate system u^i. When there is an affine connection in M, this condition can be given in a different form. To arrive at it, we compute the differential dT_k^{ij}. From (35) we have

$$dX_i^k = \sum_j X_j^k \vartheta_i^j - \sum_j \omega_j^k X_i^j. \tag{47}$$

Since Y_i^k are the elements of the inverse matrix of (X_i^j), we easily find

$$dY_i^k = -\sum_j Y_i^j \vartheta_i^j + \sum_j \omega_i^j Y_j^k. \tag{48}$$

It follows that

$$\begin{aligned} dT_k^{ij} =& -\sum_\ell T_k^{\ell j} \vartheta_\ell^i - \sum_\ell T_k^{i\ell} \vartheta_\ell^j + \sum_\ell T_\ell^{ij} \vartheta_k^\ell \\ &+ \sum_{q,r,s} Y_q^i Y_r^i X_k^s (DX_s^{qr}). \end{aligned} \tag{49}$$

Theorem 7.2.2. *In order for the function T_k^{ij} in $\mathscr{B}$ to be the form* (46) *and, therefore, to arise from a tensor field in M, it is necessary and sufficient that they should satisfy*

$$dT_k^{ij} = -\sum_\ell T_k^{\ell j} \vartheta_\ell^i - \sum_\ell T_k^{i\ell} \vartheta_\ell^j + \sum_\ell T_\ell^{ij} \vartheta_k^\ell = \sum_\ell T_{k\ell}^{ij} \vartheta^\ell. \tag{50}$$

The functions $T_{k\ell}^{ij}$ are related to the covariant differential DX_s^{qr} of the tensor field by the formula

$$\sum_\ell T_{k\ell}^{ij} \vartheta^\ell = \sum_{q,r,s} Y_q^i Y_r^j X_k^s (DX_s^{qr}). \tag{51}$$

Similar relations exist between tensor fields of general types in M and corresponding functions in $\mathscr{B}$.

The necessity of this condition was established above. To prove the sufficiency we shall show that $\sum_{i,j,k} X_i^q X_j^r Y_s^k T_k^{ij}$ are functions of u^i only. This follows from the formula

$$d\left(\sum_{i,j,k} X_i^q X_j^r Y_s^k T_k^{ij}\right) = 0. \bmod du^\ell,$$

which is easily verified, Formula (49) is then a consequence of (49).

Formulas for the exchange of order of covariant differentiation con be obtained from (49) by exterior differentiation. We omit the details here.

We shall give a geometrical interpretation of affine connection in terms of the relationship between M and $\mathscr{B}$. For this purpose we notice that there is in $\mathscr{B}$ a group of homeomorphisms defined in a natural way as follows: Denote by $pe_1 \cdots e_n$ the frames over M. We call a translation in $\mathscr{B}$ a homeomorphism $pe_1 \cdots e_n \to pe_1' \cdots e_n'$ defined by

$$e_i' = \sum_j a_i^j e_j, \tag{52}$$

where a_i^j are constants such that $(a_i^j) \neq 0$. Clearly all the translations in $\mathscr{B}$ form a group. From our general discussions on differentiable mappings of manifolds it follows that a translation T induces on isomorphism of the tangent space $V(b)$ at a point $b \in \mathscr{B}$ onto the tangent space $V(T(b))$ at the image point $T(b)$. This in turn induce a dual isomorphism of $V^*(T(b))$ onto $V^*(b)$.

Theorem 7.2.3. *Let $\pi : \mathscr{B} \to M$ be the projection of $\mathscr{B}$ onto M, which assigns to a frame $b = pe_1 \cdots e_n$ the point $p \in M$. The definition of an affine connection in M is equivalent to that of a family of linear subspaces $g(b)$ supplementary to $\pi^* V^*(p)$ in the space of covectors $V^*(b)$ at b (i.e., $g(b)$ and $\pi^* V^*(p)$ span $V^*(b)$ and have only the zero-vector in common), such that the family $g(b)$ is invariant under the group of translations in $\mathscr{B}$.*

We notice that $g(b)$ is a vector space of dimension n^2. If $\mathscr{B}'$ denotes the bundle over $\mathscr{B}$ of the n-dimensional linear subspaces of covectors, then g is a cross-section of this bundle.

The subspace $g(b)$ can be defined by n^2 linear differential forms which, together with ϑ^i, span $V^*(b)$. From (30) it follows that the forms defined in a local coordinate system by $\sum_j Y_j^k dX_i^j$ are determined mod ϑ^ℓ. Since these forms are clearly linearly independent, $g(b)$ defines and can be defined by the n^2 linear differential forms

$$\vartheta_i^k = \sum_j (dX_i^j + \varphi_i^j) Y_j^k,$$

where φ_i^j are linear combinations of du^k. Our problem is to study the forms φ_i^j when the subspaces $g(b)$ have the property of being invariant under the translations (52). It fact, denote by u^i, $x_i'^k$ the local coordinates at the image point b' of the point b under a translation (52) and denote the quantities at b' by the same symbols with dashes. Then we have

$$X_i'^k = \sum_j a_i^j X_j^k, \qquad Y_k^i = \sum_j a_j^i Y_k'^j,$$

and

$$T^* \vartheta_i'^k = \sum_j \left(\sum_\ell a_i^\ell dX_\ell^j + \varphi_i'^j \right) Y_j'^k,$$

where T^* is the dual mapping of the translation T. The invariance of $g(b)$ under T, therefore, implies that

$$\varphi_i'^j = \sum_\ell a_i^\ell \varphi_\ell^j.$$

From this we get

$$\sum_j Y_i'^j \varphi_j'^k = \sum_j Y_i^j \varphi_j^k.$$

It follows that the expression in the right-hand side of the last equation is invariant tinder translations and is, therefore, independent of X_ℓ^m. Hence we can put

$$\sum_j Y_i^j \varphi_j^k = \omega_i^k$$

or

$$\varphi_j^k = \sum_i X_j^i \omega_i^k,$$

where ω_i^k are linear differential forms in u^j only. The expressions so obtained for ϑ_i^k are identical with those in (35). Hence the family of linear subspaces $g(b)$ defines an affine connection. Our discussion also shows that an affine connection defines a family of $g(b)$ with the properties given in the theorem. Hence The theorem is proved.

7.3 Groups of holonomy

The results of this section arise when we compare a general affine connection with the affine connection of the ordinary affine space. According to the discussion at the end of §1 the latter has the property that relatively to the affine coordinates all Γ_{ij}^k are aero. We say, therefore, that an affine connection is locally flat if a local coordinate system exists relatively to which all r_{ij}^k are zero: then the torsion tensor and curvature tensor must be zero. The converse is also true:

Theorem 7.3.1. *An affine connection is locally flat if and only if both the torsion and the curvature vanish.*

We have only to prove the sufficiency. By (4) and (5) it is enough to prove that a local coordinate system u^{*i} exist such that

$$du^{*i} = \sum_k p_k^i du^k, \tag{53}$$

$$dp_u^i = \sum_j p_j^i \omega_k^j.$$

In this system we regard u^i, p_k^i, u^{*i} as variables and apply Frobenius's Theorem on completely integrable differential systems. Modulo the equations (53) we get, by exterior differentiation,

$$\sum_k dp_k^i \wedge du^k \equiv \sum_{j,k} p_j^i \omega_k^j \wedge du^k = 0,$$

$$\sum_j (dp_j^i \wedge \omega_j^i + p_j^i d\omega_k^j) \cong \sum_j p_j^i \left(\sum_\ell \omega_\ell^j \wedge \omega_k^\ell + d\omega_k^j \right) = 0$$

so that the conditions in Frobenius's Theorem are satisfied. It follows that in a neighbourhood of a point (u_0^i) in the space of the variables u^i there are functions $u^{*i}(u^k)$, $p_k^i(u^j)$ satisfying (53), such that the initial values $u^{*i}(u_0^k)$, $p_k^i(u_0^j)$ can be arbitrarily assigned. In particular, we can assign the initial values such that $|p_k^i(u_0^j)| \neq 0$. By the first equation of (53) this means that the functional determinant

$$\left|\frac{\partial u^{*i}}{\partial u^k}(u_0^j)\right| \neq 0.$$

Hence u^{*i} form a local coordinate system in a neighbourhood of (u_0^i). This proves the theorem.

We proceed to give a more geometrical proof of the theorem, which will have further implications. As discussed in the last section, an affine connection determines n^2+n linear differential forms ϑ^i, ϑ_i^j in $\mathscr{B}$, which are linearly independent. Suppose $\mathscr{B}^*$ be the principal bundle of a second of finely connected manifold M^* of dimension n, with the corresponding differential forms ϑ^{*i}, ϑ_i^{*j}. Let V and V^* be coordinate neighbourhoods in $\mathscr{B}$ and $\mathscr{B}^*$, with the local coordinates (u^i, X_i^k) and (u^{*i}, X_i^{*k}) respectively, and let $f : V \longrightarrow V^*$ be a local differentiable homeomorphism such that

$$f^*\vartheta^{*i} = \vartheta^i, \quad f^*\vartheta_i^{*j} = \vartheta_i^j, \tag{54}$$

where the differential forms denote their restrictions to the respective neighbourhoods. From the first equation we see that $\vartheta^{*i} = 0$ implies $\vartheta^i = 0$, which means that frames with the same origin in $\varphi^*(V^*)$ are mapped into frames with the same origin in $\varphi(V)$, where φ^* denotes the projection of $\mathscr{B}^*$ onto M^*. It follows that f induces a differentiable homeomorphism $f' : \varphi(V) \longrightarrow \varphi^*(V^*)$. If, by means of f', we take the local coordinates u^i of $p \in \varphi(V)$ to be those of $f'(p)$, then $q \in V$ and $f(q) \in V^*$ have the same local coordinates u^i, X_i^k. From the formulas for ϑ^i, ϑ_i^j in terms of the local coordinates u^i, X_i^k it follows that the affine connection in M^* has, relative to u^i, the same components as the affine connection in M. These considerations justify the definition:

Definition 7.3.2. Let $\mathscr{B}$ and $\mathscr{B}^*$ be the principal bundles of two of finely connected manifolds, and let V, V^* be two of their neighbourhoods. The two affine connections are called *locally equivalent* (relatively to V, V^*), if there is a differentiable homeomorphism f of V onto V^* such that equations (54) hold.

With this notion of local equivalence we can identify the locally flat affinely connected manifolds with the ones which are locally equivalent to the ordinary affine space. In fact, let M be locally flat. Denote by A^n the ordinary affine space and by $\mathscr{B}_0$ its principal bundle, which is the space of all frames in A^n. By definition there are local coordinates u^i in a neighbourhood of M relatively to which the affine connection has all components equal to 0. The mapping f which maps the point (u^i, X_i^k) of $\mathscr{B}$ into the frame A^n whose origin has the coordinates u^i and whose vectors have the components $(X_i^1, \ldots, X_i^n)$ establishes the local equivalence desired. Conversely, suppose the affine connection in M be locally equivalent to that of A^n. As local coordinates in M we can then take those of the corresponding point in A^n. From (54) it follows that relative to this local coordinate system the affine connection will have all its components equal to zero.

Since the dual homomorphism of a differentiable mapping commutes with both exterior multiplication and exterior differentiation, it follows from (54), (42) that

$$f^*\Theta^{*i} = \Theta^i, \quad f^*\Theta_i^{*j} = \Theta_i^j. \tag{55}$$

In particular, when the torsion and curvature tensors of M^* are zero, the same must be true of M:

$$\Theta^i = 0, \qquad \Theta_i^j = 0.$$

The sufficiency of the condition in Theorem 7.3.1 says that these conditions are sufficient for equations (54) to have a solution in f. This amounts to solve (locally) the differential system

$$\vartheta^{*i} = \vartheta^i, \qquad \vartheta_i^{*j} = \vartheta_i^j$$

for u^{*i}, X_i^{*k} as functions of u^i, X_i^k where both torsion and curvature tensors of M and M^* are supposed to be zero. Since the conditions in Frobenius's Theorem are satisfied, the existence of the desired solutions follows. This gives a second proof of Theorem 7.3.1.

Although the two proofs are not essentially different, this treatment has a geometrical interpretation. It means that if both torsion and curvature tensors of M are zero, the frames with origins in a neighbourhood in M can be mapped by a homeomorphism onto the frames in a neighbourhood in A^n so that equations (54) hold. The latter means intuitively that the relative positions of the frames remain unchanged.

It is obviously not possible to do the same for a general affine connection. But the above geometrical situation has in a certain sense a generalization. Let M be a manifold with an arbitrary affine connection. Let $C(t)$ a parametrized curve in M, t being the parameter. We restrict ourselves to the frames whose origins lie on $C(t)$, and we shall denote the restrictions of the differential forms ϑ^i, ϑ_i^j to this submanifold of frames by the same symbols. To describe these frames we take a particular family $pa_1 \cdots a_n(t)$ along $C(t)$, one at each point of $C(t)$. Then a general frame of the family will be $p(t)e_1 \cdots e_n$, where

$$e_i = \sum_k x_i^k a_k(t), \quad \det(x_i^k) \neq 0. \tag{56}$$

We proceed to prove that the frames $pe_1 \cdots e_n$ along $C(t)$ can be mapped into frames in A^n such that ϑ^i, ϑ_i^j are the dual images of corresponding differential forms in A^n. Geometrically this can be described by saying that we "develop" these frames into the affine space so that their relative positions remain unchanged. If $p'e_1' \cdots e_n'$ denotes the image of $pe_1 \cdots e_n$, this means that the vectors $p', e_1', \ldots, e_n'$ satisfy the differential system

$$dp' = \sum_i \vartheta^i e'_i,$$
$$de'_i = \sum_k \vartheta_i^k e'_k. \tag{57}$$

This is a differential system with variables which are t, x_i^k, and the components of the vectors $p', e'_1, \ldots, e'_n$. Since Θ^i, Θ_i^j are quadratic in the differentials of the local coordinates, they vanish along a parametrized curve. Just as in the locally flat case, it follows that the system (57) is completely integrable. Therefore, there is one and only one family of vectors $p', e'_1, \ldots, e'_n$ satisfying (57) and taking arbitrary initial positions for $t = t_0$, $x_i^k = \delta_i^k$.

We shall first prove that $p'e'_1 \cdots e'_n$ is a frame, that is, that the vectors $e'_1, \cdots, e'_n$ are linearly independent, if the same is true of the initial position. In fact, let

$$\Delta = (e'_1, \ldots, e'_n) \tag{58}$$

denote the determinant whose ith row consists of the components of e'_i. We find

$$d\Delta = \sum_i (e'_1, \ldots, e'_{i-1}, \vartheta_i^i e'_i, e'_{i+1}, \ldots, e'_n)$$
$$= \left(\sum_i \vartheta_i^i\right) \Delta.$$

Integration gives

$$\Delta = \Delta_o e^{\int \sum_i \vartheta_i^i}, \tag{59}$$

where Δ_0 denotes the initial value of the determinant. Hence Δ never vanishes if $\Delta_0 \neq 0$.

Moreover, since ϑ^i are multiples of dt, p' is a function of t only, so that its locus is a parametrized curve $C'(t)$ in A^n. Let $p'a'_1 \cdots a'_n$ be the image of $pa_1 \cdots a_n$. We wish first to determine the form ϑ_i^j mod dt. Using a local coordinate system u^i we can write

$$a_i(t) = \sum_j a_i^j(t) \frac{\partial}{\partial u^j}. \tag{60}$$

Let y_i^j be the elements of the matrix inverse to (x_j^k), so that

$$\sum_j x_i^j y_j^k = \sum_j y_i^j x_j^k = \delta_i^k.$$

From (35) we find

$$\vartheta_i^k \equiv \sum_j Y_j^k dX_i^j \equiv \sum_j y_j^k dX_i^j, \quad \text{mod } dt. \tag{61}$$

This relation and (57) imply that

$$d\left(\sum_j y_i^j e'_j\right) \equiv \sum_j \left(dy_i^j + \sum_{k,\ell} y_i^k dx_k^\ell y_\ell^j\right) e'_j = 0, \quad \text{mod } dt.$$

It follows that the vectors $\sum_j y_i^j e'_j$ are functions of t only. Hence they remain unchanged when we put $y_i^j = \delta_i^j$. But this means that

$$\sum_j y_i^j e'_j = a'_i(t) \quad \text{or} \quad e'_i = \sum_j x_i^j a'_j(t). \tag{62}$$

This shows that the vectors e'_i satisfying (57) are of a very special form.

These considerations lead us to define an affine transformation of the tangent space $V(p(t_1))$ onto the tangent space $V(p(t_2))$, where $p(t_1)$ and $p(t_2)$ are two points of $C(t)$ corresponding to the parameters t_1 and t_2 respectively. In fact, a point of $V(p(t))$ can be defined by $\sum_i x^i a_i(t)$. We map the point $\sum_i x^i a_i(t_1)$ of $V(p(t_1))$ into the point $\sum_i x^{*i} a_i(t_2)$ of $V(p(t_2))$, where x^{*i} are defined by the equation

$$p'(t_1) + \sum_i x^i a'_i(t_1) = p'(t_2) + \sum_i x^{*i} a'_i(t_2). \tag{63}$$

This condition is the analytical expression of the classical geometrical operation of "developing" the tangent spaces along a curve into an affine space and comparing these tangent space by means of the development. We shall denote this affine transformation by $T_{t_2t_1}$. Notice that $T_{t_2t_1}$ is in general not a linear mapping of the tangent spaces, because the image of $p(t_1)$ may not be $p(t_2)$ (that is, $x^i = 0$ does not necessarily imply $x^{*i} = 0$). There is, however, a linear mapping $T'_{t_2t_1}$ of the vector space $V(p(t_1))$ into $V(p(t_2))$ by which the vector $\sum_i x^i a_i(t_1)$ goes into the vector $\sum_i x^{*i} a_i(t_2)$, where x^{*i} are defined by the condition

$$\sum_i x^1 a'_i(t_1) = \sum_i x^{*i} a'_i(t_2). \tag{64}$$

In order for the mappings $T_{t_2t_1}$ and $T'_{t_2t_1}$ to be meaningful they must be independent of the following choices which have been made

1) Choice of the initial frame $p'e'_1 \cdots e'_n$ for $t = t_0$, $x_i^k = \delta_i^k$;
2) Choice of the family of frames $pa_1 \cdots a_n(t)$ along $C(t)$.

To prove their independence from the choice 1) we denote by p'^k, $e_i'^k$ the components of p', e'_i respectively. Then equations (57) can be written

$$\begin{aligned} dp'^k &= \sum_i \vartheta^i e_i'^k, \\ de_i'^k &= \sum_j \vartheta_i^j e_j'^k. \end{aligned} \tag{65}$$

It follows that if p'^k, $e_i'^k$ are solutions of (65), the functions

$$p''^k = h^k + \sum_{\ell} g_\ell^k p'^k,$$
$$e_i''^k = \sum g_\ell^k e_i'^\ell, \quad \det(g_q^k) \neq 0, \tag{66}$$

where g_ℓ^k, h^k are constants, are also solutions of (65). There are, moreover, the most general solutions, since, by proper choice of the constants g_ℓ^k, h^k, the solutions p''^k, $e_i''^k$ can take arbitrary initial values for $t = t_0$ and $x_i^k = \delta_i^k$. Obviously the conditions (63) and (64) remain unchanged when p', e_i' are replaced by p'', e_i''. This proves the independence of the mappings $T_{t_2t_1}$ and $T'_{t_2t_1}$ from the choice 1).

Independence of the choice 2) follows immediately from equations (56) and (62), when x_i^k are taken as functions of t.

The mappings $T_{t_2t_1}$ and $T'_{t_2t_1}$ have clearly the following properties:

$$T_{t_3t_1} = T_{t_3t_2}T_{t_2t_1}, \quad T_{t_2t_1} = T_{t_1t_2}^{-1},$$
$$T'_{t_3t_1} = T'_{t_3t_2}T'_{t_2t_1}, \quad T'_{t_2t_1} = T_{t_1t_2}'^{-1} \tag{67}$$

where t_1, t_2, t_3 are three values for t.

With these preparations we can define some important invariants of an affine connection, namely, the *groups of holonomy*. Let M be connected, and let O be a point of M. Let $\{\gamma\}$ be the family of closed parametrized curves in M with 0 as initial point. To each γ there corresponds affine transformation T_γ of the tangent space V at 0. All these transformations clearly form a group, we shall call the *group of holonomy at* 0 and shall denote by H_0. Similarly, the linear mappings T'_γ corresponding to the curves γ also form a group, to be called the *restricted group of holonomy at O and* denoted by H'_0.

Suppose that, instead of 0, we choose a point $0'$ as initial point. Let β be a parametrized curve joining 0 to $0'$. By means of β we map a closed parametrized curve γ through 0 into $\gamma' = \beta\gamma\beta^{-1}$. This induces a mapping of T_γ into $T_{\gamma'}$ and defines a homomorphism of H_0 into $H_{0'}$. It is easy to see that this homomorphism is an isomorphism onto. In particular, it follows that the group of holonomy, as an abstract group, is independent of the choice of 0. The same result holds for the restricted group of holonomy.

Definition 7.3.3. An affine connection is called a *Weyl connection*, if its restricted group of holonomy leaves invariant an equation $Q = 0$, where Q is a positive definite quadratic form. It is called a matrical connection, if its restricted group of holonomy leaves invariant a positive definite quadratic form.

The importance of the groups of holonomy lies in the fact that notions which are invariant under them can be defined. To illustrate this, suppose the affine connection to be metrical connection. We shall show that in this case we can define scalar product of two tangent vectors with the same origin. In fact, let the scalar product of two vectors at a point O be defined. Suppose $p \in M$ and v_1, v_2 two vectors at p. We join 0 and p by a parametrized curve α, and denote by T'_α the corresponding linear mapping of $V(0)$ onto $V(p)$. Then we define the scalar product as

$$v_1 \cdot v_2 = (T_\alpha'^{-1} v_1) \cdot (T_\alpha'^{-1} v_2). \tag{68}$$

To prove that this is independent of the choice of α, let β be a second parametrized curve joining 0 to p, and T'_β the corresponding linear mapping of $V(0)$ onto $V(p)$. Then $T' = T_\beta'^{-1} T'_\alpha$ is an element of H'_0 and leaves the scalar product of vectors at 0 invariant. Hence we have

$$\begin{aligned}(T_\alpha'^{-1} v_1) \cdot (T_\alpha'^{-1} v_2) &= (T' T_\alpha'^{-1} v_1) \cdot (T' T_\alpha'^{-1} v_2) \\ &= (T_\beta'^{-1} v_1) \cdot (T_\beta'^{-1} v_2),\end{aligned}$$

which shows that the definition (68) is independent of the choice of α.

7.4 Affine normal coordinates

Normal coordinates are local coordinates at a point 0 relative to which the paths through 0 will be defined by linear equations. They are, therefore useful in the study of local properties of affine connections. It is possible to establish the existence of normal coordinates in a more general form so as to allow 0 to vary in a neighbourhood, as we shall show presently.

We recall that in a neighbourhood with the local coordinates u^i the paths of an affine connection are defined by the system of differential equations

$$\frac{d^2u^i}{dt^2} + \sum_{j,k} \Gamma^i_{jk} \frac{du^j}{dt}\frac{du^k}{dt} = 0. \tag{21}$$

Let

$$u^i = f^i(t, x^k, \alpha^k) \tag{69}$$

be the equations of the path satisfying the initial condition

$$u^i(0) = x^i, \quad \left(\frac{du^i}{dt}\right)_{t=0} = \alpha^i. \tag{70}$$

The functions $f^i(t, x^k, \alpha^k)$ are of class ∞ in their arguments. Since they remain solutions when t is multiplied by a non-zero constant factor, we have

$$f^i(t, x^k, \alpha^k) = F^i(x^k, \alpha^k t). \tag{71}$$

we define

$$y^i = f^i(1, x^k, \alpha^k) = F^i(x^k, \alpha^k). \tag{72}$$

Considering x^i, y^i as functions of x^k, α^k, we see that, at $t = 0$, their functional determinant is equal to 1. It follows that there is a neighbourhood U in M such that in a neighbourhood of the diagonal of the topological product $U \times U$, the variables x^k, α^k form an admissible coordinate system. The α^k are called the *normal coordinates at the point p whose local coordinates are* x^k. In terms of the normal coordinates α^k the paths through p are given by equations of the form

$$\alpha^i = \beta^i t, \tag{73}$$

where β^i are constants. In particular, when the point p is fixed, it has a neighbourhood in which the normal coordinates α^i are admissible such that the paths through p are given by the equations (73). As a consequence, the point p has a neighbourhood U in which every point can be joined to p by exactly one path belonging to U.

For later use we state the following lemma, which follows immediately from the existence of normal coordinates.

Lemma 7.4.1. *Every point O of M has a neighbourhood V such that any point $p \in V$ has a coordinate neighbourhood covered by normal coordinates at p, which contains V.*

Throughout the rest of this section we suppose M to be without torsion. Consider normal coordinates at a fixed point 0 of M. We take a frame at 0 and displace it parallelly along the paths through 0, getting thus a family of frames in a neighbourhood of 0. Denote by ϑ^i, ϑ_i^j the restrictions of these differential forms to this family of frames. Then ϑ^i, ϑ_i^j are differential forms in t, α^k. When α^k remain constant, the frames are parallel to themselves. It follows that we can write

$$\vartheta^i = \alpha^i dt + \overline{\vartheta}^i, \tag{74}$$

$$\vartheta_j^i = \overline{\vartheta}_j^i, \tag{75}$$

where $\overline{\vartheta}^i$, $\overline{\vartheta}_i^j$ do not involve dt. These differential forms satisfy the fundamental equations (42). Substituting them into (42) and retaining only the terms which involve dt, we get

$$-dt \wedge d\alpha^i + dt \wedge \frac{\partial \overline{\vartheta}^i}{\partial t} = \sum_j \alpha^j dt \wedge \overline{\vartheta}_j^i,$$

$$dt \wedge \frac{\partial \overline{\vartheta}_i^j}{\partial t} = \sum_{k,\ell} S_{ik\ell}^j \alpha^k dt \wedge \overline{\vartheta}^\ell,$$

where the operator $\frac{\partial}{\partial t}$ applied on a differential form denotes partial differentiation with respect to t of the coefficients. From these equations it follows that

$$\frac{\partial \overline{\vartheta}^i}{\partial t} = d\alpha^i + \sum_j \alpha^j \overline{\vartheta}^i_j, \tag{76}$$

$$\frac{\partial \overline{\vartheta}^j_i}{\partial t} = \sum_{k,\ell} S^j_{ik\ell} \alpha^k \overline{\vartheta}^\ell. \tag{77}$$

These give in turn the following partial differential equation satisfied by $\overline{\vartheta}^i$:

$$\frac{\partial^2 \overline{\vartheta}^i}{\partial t^2} = \sum_{j,k,\ell} S^i_{jk\ell} \alpha^j \alpha^k \overline{\vartheta}^\ell. \tag{78}$$

Since we have, at 0,

$$\overline{\vartheta}^i = 0, \qquad \overline{\vartheta}^j_i = 0, \tag{79}$$

the $\overline{\vartheta}^i$ satisfy the initial conditions

$$(\overline{\vartheta}^i)_0 = 0, \qquad \left(\frac{\partial \overline{\vartheta}^i}{\partial t}\right)_0 = d\alpha^i. \tag{80}$$

Relations (78) and (80) determine $\overline{\vartheta}^i$ completely.

Theorem 7.4.2. *An affine connection without torsion is locally determined by the curvature tensor. If the manifold and the connection are both analytic, the affine connection is locally determined by the values of the components of the curvature tensor and its successive covariant differentials at a point.*

This theorem shows the important role played by the curvature tensor in the local properties of an affine connection. The second part gives the degree of arbitrariness by which an affine connection is determined.

To prove the first part of the theorem we notice that when the curvature tensor is given, $S^j_{ik\ell}$ are known functions of t, α^k. Then the solutions $\overline{\vartheta}^i$, $\overline{\vartheta}^j_i$ of (76) are uniquely determined by the initial conditions (79). By (74) they inturn determine ϑ^i, ϑ^j_i, and hence the affine connection.

Suppose analyticity assumptions be new made. The functions $S^j_{ik\ell}$ will then be determined by the values at 0 of $S^j_{ik\ell}$ and $\frac{\partial^\lambda}{\partial t^\lambda} S^j_{ik\ell}$, $\lambda = 1, 2, \ldots$. Using the analogue of (50), we find, when α^k are constants,

$$dS^j_{ik\ell} = \sum_q S^j_{ik\ell,q} \alpha^{qdt}.$$

This gives

$$\frac{\partial S^j_{ik\ell}}{\partial t} = \sum_q S^j_{ik\ell,q} \alpha^q.$$

More generally, we get

$$\frac{\partial^\lambda S^j_{ik}}{\partial t^\lambda} = \sum_{q_1,\ldots,q_\lambda} S^j_{ik\ell,q_1,\ldots,q_\lambda} \alpha^{q_1} \cdots \alpha^{q_\lambda}. \tag{81}$$

This proves the second statement of the theorem.

8

Riemannian Manifolds

Let M be a differentiable manifold of dimension n and class ∞. A *Riemann metric* in M is a field of positive definite symmetric covariant tensors of order 2. This tensor field will be called the fundamental tensor field or simply the *fundamental tensor*. Then a Riemann metric is given in M, M is called a Riemannian manifold. The object of this chapter is to develop the elementary concepts of Riemannian manifolds.

First of all, we define the scalar product of two vectors with the same origin as the contraction of their tensor product with the fundamental tensor at the point. If g_{ij} are the components of the fundamental tensor with respect to the natural frame of a local coordinate system u^i and X^i, Y^i are those of the vectors x, y, the scalar product is given by

$$x \cdot y = \sum_{i,j} g_{ij} X^i Y^j. \tag{1}$$

The *length* of a vector x is the positive square root of the product $x^2 = x \cdot x$.

The Riemann metric also enables us to define the arc length of a parametrized curve. For the quadratic differential form

$$ds^2 = \sum_{i,j} g_{ij} du^i du^j \tag{2}$$

is independent of the local coordinates u^i. If $u^i = u^i(t)$ defines a parametrized curve, the integral

$$s = \int_{t_0}^{t_1} \sqrt{\sum_{i,j} g_{ij} \frac{du^i}{dt} \frac{du^j}{dt}}\, dt \tag{3}$$

defines the arc length between the points $t = t_0, t_1$.

It may be remarked that every manifold has a Riemann metric. To prove this, take a locally finite covering $\{U_i\}$ of M by coordinate neighbourhoods. For each U_i let ds_i^2 be a positive definite quadratic differential form defined in it. Let g_α be a corresponding partition of unity according to Theorem 6.5.5. Then $\sum_\alpha g_\alpha ds_i^2$, where i is an index such that U_i contains the carrier of g_α, defines a Riemann metric in M.

One of the main results of Riemannian Geometry is the discovery of the predominant role played by a certain affine connection: the parallelism of Levi-Civita.

8.1 The parallelism of Levi-Civita

Theorem 8.1.1. *On a Riemannian manifold there is one affine connection with the following properties:*

1) *Its torsion tensor is zero;*
2) *The scalar product of two vectors remains unchanged, when they are displaced parallelly along a curve.*

To prove the theorem we use a local coordinate system u^i. Consider a parametrized curve $u^i = u^i(t)$, and along it the tangent vectors with the components $X^i(t)$, $Y^i(t)$. By Definition 7.1.2, the conditions that these vectors are parallel along the curve are

$$\begin{aligned} \frac{dX^i}{dt} + \sum_{j,k} \Gamma^i_{jk} X^k \frac{du^j}{dt} &= 0, \\ \frac{dY^i}{dt} + \sum_{j,k} \Gamma^i_{jk} Y^k \frac{du^j}{dt} &= 0. \end{aligned} \tag{4}$$

Property 2) is equivalent to saying that

$$\frac{d}{dt}\left(\sum_{i,j} g_{ij} X^i Y^j\right) = 0$$

whenever these conditions are satisfied. Using (4) we find

$$\frac{d}{dt}\left(\sum_{i,j} g_{ij} X^i Y^i\right) = \sum_{i,j}\left(\frac{dg_{ij}}{dt} - \sum_{k,\ell} g_{i\ell}\Gamma^\ell_{jk}\frac{du^k}{dt} - \sum_{k,\ell} g_{j\ell}\Gamma^\ell_i \frac{du^k}{dt}\right) X^i Y^j.$$

Since this holds for all parametrized curves and all vectors $X^i(t)$, $Y^i(t)$, we get

$$\frac{\partial g_{ij}}{\partial u^k} = \Gamma_{kij} + \Gamma_{kji} \tag{5}$$

where we put

$$\Gamma_{kij} = \sum_{\ell} g_{\ell i} \Gamma^\ell_{kj}. \tag{6}$$

On the other hand, the vanishing of the torsion tensor can be expressed by

$$\Gamma^i_{kj} = \Gamma^i_{jk} \quad \text{or} \quad \Gamma_{kij} = \Gamma_{jik}. \tag{7}$$

Permuting equations (5) cyclically in i, j, k, we get

$$\frac{\partial g_{jk}}{\partial u^i} = \Gamma_{ijk} + \Gamma_{ikj},$$
$$\frac{\partial g_{ki}}{\partial u^j} = \Gamma_{jki} + \Gamma_{jik}.$$

Since Γ_{ijk} is symmetric in the first and third indices, these two equations, together with (5), give

$$\Gamma_{ijk} = \frac{1}{2}\left(\frac{\partial g_{ij}}{\partial u^k} + \frac{\partial g_{jk}}{\partial u^i} - \frac{\partial g_{ik}}{\partial u^j}\right) \tag{8}$$

or

$$\Gamma^j_{ik} = \frac{1}{2}\sum_\ell g^{j\ell}\left(\frac{\partial g_{i\ell}}{\partial u^k} + \frac{\partial g_{\ell k}}{\partial u^i} - \frac{\partial g_{ik}}{\partial u^\ell}\right), \tag{9}$$

when g^{ij} are defined by

$$\sum_j g_{ij} g^{jk} = \delta^k_i. \tag{10}$$

Hence the affine connection is completely determined. The components Γ_{ijk} and Γ^j_{ik} are called respectively the *Christoffel symbols of the first and second kind.*

The affine connection which is so defined is called the *connection of Levi-Civita*, and the parallelism of vectors arising from it, the *parallelism of Levi-Civita*. It is clear that its restricted group of holonomy is a subgroup of the group which leaves the scalar product invariant.

It follows that the connection of Levi-Civita is a metrical connection. For such connections it is important to consider, instead of the bundle $\mathscr{B}$, the bundle $\mathscr{Q}$ of rectangular frames, that is, frames $pe_1 \cdots e_n$ which satisfy the condition that $e_1, \ldots, e_n$ are mutually perpendicular unit vectors. Clearly $\mathscr{Q}$ is a closed submanifold of $\mathscr{B}$; its dimension is $n(n+1)/2$. Differential forms and functions in $\mathscr{B}$ will have restrictions to $\mathscr{Q}$. Their properties are given by the following theorem.

Theorem 8.1.2. *Let ϑ_i, ϑ_{ij}, Θ_{ij}, $S_{ijk\ell}$ be the restrictions to $\mathscr{Q}$ of ϑ^i, ϑ^j_i, Θ^j_i, $S^j_{ik\ell}$ respectively. Then*

$$\vartheta_{ij} + \vartheta_{ji} = 0, \qquad \Theta_{ij} + \Theta_{ji} = 0, \tag{11}$$

and the forms ϑ^i, ϑ_{ij}, $i < j$, are linearly independent. These forms satisfy the equations

$$\begin{cases} d\vartheta_i = \sum_j \vartheta_j \wedge \vartheta_{ji}, \\ d\vartheta_{ij} = \sum_k \vartheta_{ik} \wedge \vartheta_{kj} + \Theta_{ij}, \end{cases} \tag{12}$$

$$d\Theta_{ij} + \sum_k \Theta_{ik} \wedge \vartheta_{kj} - \sum_k \vartheta_{ik} \wedge \Theta_{kj} = 0, \tag{13}$$

and the functions $S_{ijk\ell}$ have the following symmetry properties:

$$S_{ijk\ell} = -S_{jik\ell} = -S_{ij\ell k}, \tag{14}$$

$$S_{ijk\ell} = S_{k\ell ij}, \tag{15}$$

$$S_{ijk\ell} + S_{ik\ell j} + S_{i\ell jk} = 0. \tag{16}$$

If, in terms of a local coordinate system u^i,

$$\ell_i = \sum_k X_i^k \frac{\partial}{\partial u^k},$$

the functions $S_{ijk\ell}$ are given by

$$S_{ijk\ell} = \sum_{p,q,r,s} X_i^p X_j^q X_k^r X_\ell^s R_{pqrs}, \tag{17}$$

where

$$R_{pqrs} = \sum_i g_{qi} R^i_{prs}. \tag{18}$$

To prove the theorem we apply the process of §7.3, by developing the frame along a parametrized curve C into the affine space A^n. The image frames $p'e'_1 \cdots e'_n$ then satisfy the differential system

$$dp' = \sum_i \vartheta^i e'_i, \qquad de'_i = \sum_j \vartheta_i^j e'_j.$$

For a particular value t_0 of the parameter let v, w be tangent vectors at $p(t_e)$, and let v', w' be their images in the tangent space $V(p'(t_0))$. We define the scalar product

$$v' \cdot w' = v \cdot w. \tag{19}$$

By identifying A^n with one of its tangent spaces, we introduce thus a scalar product in A^n. Since Levi-Civita's connection preserves scalar product, the scalar product so introduced in A^n is independent of the choice of t_0. It follows that we can develop in A^n the restangular frames along C, whose images $p'e'_1 \cdots e'_n$ satisfy the differential system

$$dp' = \sum_i \vartheta_i e'_i, \qquad de'_i = \sum_j \vartheta_{ij} e'_j, \tag{20}$$

with

$$e'_i \cdot e'_j = \delta_{ij}. \tag{21}$$

Differentiating the last equation, we get

$$0 = d(e'_i \cdot e'_j) = \vartheta_{ij} + \vartheta_{ji}.$$

Since this is true along any parametrized curve C, it is true in general. This proves the first equation in (11). Equations (12), (13) are restrictions to $\mathcal{Q}$ of (42), (43) respectively, accounts being taken of the fact that our connection is without torsion. The second equation of (11) is then a consequence of (12) and the first equation of (11).

It is clear that the ϑ_i are linearly independent. On the other hand, the forms ϑ_{ij}, $i < j$, are linearly independent, modulo the ϑ_k, since $\mathcal{Q}$ contains all rectangular frames at p, for any fixed p. It follows that forms ϑ_i, ϑ_{ij}, $i < j$, are linearly independent. Since

$$\Theta_{ij} = \frac{1}{2} \sum_{k,\ell} S_{ijk\ell} \vartheta_k \wedge \vartheta_\ell, \tag{22}$$

equation (14) follows from the second equation of (11) and the fact that ϑ_i are linearly independent.

Exterior differentiation of the first equation of (12) gives, when we make use of the equations (12) themselves,

$$\sum_j \vartheta_j \wedge \Theta_{ji} = 0. \tag{23}$$

By substituting into it the expressions for Θ_{ji} and equating to zero the coefficient of a general term, we get (16). It is easy to derive (15) as a consequence of (14) and (16).

To prove (17), which expresses the $S_{ijk\ell}$ as functions of the components of the vectors of a rectangular frame in terms of a natural base, we start from the relation (Chapter 7, (35), (38), (15))

$$\Theta_i^j = \frac{1}{2} \sum_{k,\ell,p,q} Y_k^j X_i^\ell R_{\ell pq}^k du^p \wedge du^q.$$

Since the frames are rectangular, we have

$$\sum_{k,\ell} X_i^k X_j^\ell g_{k,\ell} = \delta_{ij},$$

or

$$\sum_k Y_k^i g^{k\ell} = \sum_k \delta^{ik} X_k^\ell.$$

It follows that

$$\sum_k Y_k^j R_{\ell pq}^k = \sum_{i,m} \delta^{ij} X_i^m R_{\ell mpq},$$

whence

$$\Theta_i^j = \frac{1}{2} \sum_{r,\ell,m,p,q} \delta^{rj} X_i^\ell X_r^m R_{\ell mpq} du^p \wedge du^q$$
$$= \frac{1}{2} \sum_{\substack{\ell,m,p,q \\ r,s,t}} \delta^{rj} X_i^\ell X_r^m X_s^p X_t^q R_{\ell mpq} \vartheta^s \wedge \vartheta^t.$$

From this equation we get (17). Thus all the statements of Theorem 8.1.2 are proved.

Remark 8.1.3. *From* (17) *we see that the functions* $R_{ijk\ell}$ *satisfy the same symmetry relations* (14), (15), (16) *as* $S_{ijk\ell}$.

The importance of the connection of Levi-Civita lies in the fact that it solves the following "equivalence problem" in local Riemannian Geometry: Given two Riemann metrics in the local coordinates u^i and u^{*i} respectively, determine whether there is a local differentiable homemomorphism carrying one metric to the other.

Suppose, in fact, that such a homeomorphism f exists, carrying a neighbourhood U of one manifold into a neighbourhood U^* of the second, such that $f * ds^{*2} = ds^2$. f induces a homeomorphism of the corresponding tangent bundles and, being an isomorphism on the vector spaces at corresponding points, also induces a homeomorphism of the corresponding principal bundles. Denote by $\mathscr{Q}$ and $\mathscr{Q}^*$ the bundles of rectangular frames over U and U^* respectively. Since the induced homeomorphism of f preserves the scalar product of vectors with the same origin, f induces a homeomorphism

$$\mathscr{F} : \mathscr{Q} \to \mathscr{Q}^*. \tag{24}$$

We proved above that the connection of Levi-Civita is uniquely determined by the Riemann metric and that it in turn determines the forms ϑ_i, ϑ_{ij} in $\mathscr{Q}$. It follows that

$$\mathscr{F}^* \vartheta_i^* = \vartheta_i, \quad \mathscr{F}^* \vartheta_{ij}^* = \vartheta_{ij}. \tag{25}$$

From (12) we have

$$\mathscr{F}^* S_{ijk\ell}^* = S_{ijk\ell}. \tag{26}$$

By taking successive covariant differentials, we see at $\mathscr{F}$ must satisfy the condition

$$\mathscr{F}^* S_{ikj\ell,m}^* = S_{ijk\ell,m}, \; \mathscr{F}^* S_{ijk\ell,mp}^* = S_{ijk\ell,mp}, \cdots \tag{27}$$

Such are, therefore, the conditions on f in order that it establishes a local equivalence of two Riemann metrics. It can be proved that is sufficient to consider a finite number of the conditions (27). Moreover important is, however, the property expressed by the equations (25), which shows that the forms ϑ_i, ϑ_{ij} are "intrinsically" associated to the Riemann metric. They therefore account for the important properties of Riemannian Geometry.

Conversely, suppose a local homeomorphism f exists, satisfying (25). By the first equation of (25) it induces a homeomorphism f between U and U^*. Moreover, we have

$$f^*(ds^{*2}) = \mathscr{F}^*(\vartheta_1^{*2} + \cdots + \vartheta_n^{*2})$$
$$= \vartheta_1^2 + \cdots + \vartheta_n^2 = ds^2,$$

so that f establishes an equivalence of the two Riemann metrics.

8.2 Sectional curvature

The discussions of the last section justify analytically the importance of the quantities $S_{ijk\ell}$ or, what is the same, the Riemann-Christoffel curvature tensor. The aim of this section is to give geometrical interpretations and to derive some of their properties from the geometrical viewpoint.

Consider first the two-dimensional case. Equations (12) can be written

$$d\vartheta_1 = \vartheta_2 \wedge \vartheta_{21}, \; d\vartheta_2 = \vartheta_1 \wedge \vartheta_{12}, \qquad (28)$$
$$d\vartheta_{12} = S_{12,12}\vartheta_1 \wedge \vartheta_2.$$

It is proved in elementary differential geometry that S_{1212} is a scalar function and is equal to the negative of the Gaussian curvature.

We shall make use of this fact to study the geometrical properties of an n-dimensional Riemann manifold M at a point O. In a neighbourhood of O consider a surface (i a two-dimensional submanifold) S through O. We restrict ourselves to the family F of rectangular frames $pe_1 \cdots e_n$ having the properties: 1) p lies on S; 2) e_1 and e_2 are tangent to S at p. We denote by ϑ_i, ϑ_{ij} the restrictions of the same differential forms to this sub-family. For consistency let us agree on the following ranges of indices:

$$1 \le \alpha, \qquad \beta \le 2, \qquad 3 \le r \le n. \qquad (29)$$

To study the properties of this family of frames let u^i be local coordinates at O, and let X_i^k be the components of e_i relative to the natural base. Then we have

$$du^i = X_1^i \varphi^1 + X_2^i \varphi^2, \qquad (30)$$

where φ^1, φ^2 are linearly independent linear differential form. In particular, it follows that

$$\vartheta^r = \sum_k Y_k^r du^k = 0,$$

and that ϑ^1, ϑ^2 are linearly independent. Since we restrict ourselves to the rectangular frames, we have adopted the convention of using consistently the subscripts, and we can write (31)

$$\vartheta_r = 0. \qquad (31)$$

Taking exterior derivatives of these equations and making use of (12) §8.1 and these equations themselves, we get

$$\sum_{\alpha} \vartheta_{\alpha} \wedge \vartheta_{\alpha r} = 0.$$

Since ϑ_{α} are linearly independent, we can write

$$\vartheta_{\alpha r} = \sum_{\beta} A_{r\alpha\beta} \alpha_{\beta}, \tag{32}$$

where

$$A_{r\alpha\beta} = A_{r\beta\alpha}. \tag{33}$$

The surface S is said to be *geodesic at O*, if contains every geodesic through O and tangent to a vector in the plane Oe_1e_2. When the plane is given, such a surface exists in a neighbourhood of O, being generated by the geodesics in question. We proceed to find the condition that S is geodesic at O. For this purpose consider a parametrized curve C through O on S, which is tangent at O to the vector $\sum_{\alpha} x_{\alpha} e_{\alpha}$. When the frames along C are developed in Euclidean space, we have

$$d\left(\sum_{\alpha} x_{\alpha} e'_{\alpha}\right) = \sum_{\alpha,r} x_{\alpha} \vartheta_{\alpha\gamma} e'_r + (\cdots) e'_{\beta}.$$

It follows that C is a geodesic only if

$$\sum_{\alpha} x_{\alpha} \vartheta_{\alpha r} = 0.$$

Since this holds for arbitrary initial values of x_{α}, we have

$$(\vartheta_{\alpha r})_O = 0. \tag{34}$$

These are, therefore, necessary conditions that S be geodesic at O.

Suppose now that (34) holds at O. Using (12), we get, on restricting ourselves to the family of frames F,

$$\begin{aligned}
d\vartheta_1 &= \vartheta_a \wedge \vartheta_{21} + \sum_r \vartheta_r \wedge \vartheta_{r1} = \vartheta_2 \wedge \vartheta_{21}, \\
d\vartheta_2 &= \vartheta_1 \wedge \vartheta_{12} + \sum_r \vartheta_r \wedge \vartheta_{r2} = \vartheta_1 \wedge \vartheta_{12}, \\
d\vartheta_{12} &= \sum_r \vartheta_{1r} \wedge \vartheta_{r2} + S_{12,12} \vartheta_1 \wedge \vartheta_2.
\end{aligned} \tag{35}$$

Comparing these equations with (28), we reach the conclusion: *At any point O of Riemann manifold, the quantity $S_{12,12}$ is equal to the Gaussian curvature at O of the surface geodesic at O and tangent to the plane spanned by the first two vectors.*

Definition 8.2.1. Let p be a point of a Riemann manifold and E a plane in the tangent space at p. The Gaussian curvature at p of the surface geodesic at p and tangent to E is called the *Riemannian curvature* or the *sectional curvature* at (p,E). It will be denoted by $R(p,E)$.

Let u^i be a local coordinate system, and X^i, Y^i the components of two mutually perpendicular unit vectors which span E. The above conclusion, together with (17), gives the formula

$$R(p,E) = -\sum_{i,j,k,\ell} R_{ijk\ell} X^i Y^j X^k Y^\ell.$$

It will, however, be useful to derive a formula for $R(p,E)$ in terms of the components of any two vectors spanning E. For this purpose let X^*, Y^* be two such vectors, so that we have

$$X = aX^* + bY^*,\ Y = cX^* + dY^*,\ ad - bc \neq 0.$$

Then

$$R(p,E) = -(ab - bc)^2 \sum_{i,j,k,\ell} R_{ijk\ell} X^{*i} Y^{*j} X^{*k} Y^{*\ell}.$$

The reciprocal of the quantity $ad - bc$ is equal to the measure of the bivector spanned by the vectors X^*, Y^*, that is, the oriented area of the parallelogram with X^*, Y^* as sides. By dropping the asterisks and denoting by $m(p,X,Y)$ the measure of the bevect spanned by X, Y, we get

$$R(p,E) = -\frac{\sum_{i,j,k,\ell} R_{ijk\ell} X^i Y^j X^k Y^\ell}{m^2(p,X,Y)}. \tag{36}$$

We note in passing the following formula for the measure of a bivector:

$$m^2(p,X,Y) = \sum_{i,j,k,\ell} (g_{ik} g_{j\ell} - g_{i\ell} g_{jk}) X^i Y^j X^k Y^\ell. \tag{36a}$$

The significance of the sectional curvature is further justified by the following theorem.

Theorem 8.2.2. *The Riemann-Christoffel tensor at a point p is determined by the values of the sectional curvature for all planes through p, if the components of the tensor g_{ij} are given at p.*

Let $\overline{R}_{ijk\ell}$ be the components of a second Riemann-Christoffel tensor relative to which the sectional curvature has the same values for all planes. Then we have

$$\sum_{i,j,k,\ell} R_{ijk\ell} X^i Y^j X^k Y^\ell = \sum_{i,j,k,\ell} \overline{R}_{ijk\ell} X^i Y^j X^k Y^\ell,$$

or

$$\sum_{i,j,k,\ell} a_{ijk\ell} X^i Y^j X^k Y^\ell = 0$$

for arbitrary X^i, Y^i, where

$$a_{ijk\ell} = R_{ijk\ell} - \overline{R}_{ijk\ell}.$$

The quantities $a_{ijk\ell}$ have the symmetry properties (14), (15), (16). Supposing that the indices take different values, we get

$$\begin{aligned} &a_{ijij} = 0, \\ &a_{ijik} + a_{ikij} = 0, \\ &a_{ijk\ell} + a_{kji\ell} + a_{i\ell kj} + a_{k\ell ij} = 0. \end{aligned}$$

The second equation gives, on account of (15),

$$a_{ijik} = 0.$$

The third equation gives

$$a_{ijk\ell} = a_{i\ell jk} = a_{ik\ell j}.$$

By (16) their sum is zero, whence their common value is also zero. Thus we have proved that $a_{ijk\ell} = 0$ or

$$R_{ijk\ell} = \overline{R}_{ijk\ell}$$

for all indices i, j, k, ℓ.

Definition 8.2.3. A Riemann manifold is said to be *isotropic* at point p, if the sectional curvature in the plane element (p,E) is independent of E.

Let M be isotropic at p, with the common sectional curvature $K(p)$. By Theorem 8.2.2 it follows that a necessary and sufficient condition for this property is

$$R_{ijk\ell} = K(g_{i\ell}g_{jk} - g_{ik}g_{j\ell}). \tag{37}$$

From (17) we get

$$\begin{aligned} S_{ijk\ell} &= K \sum_{p,q,r,s} X_i^p X_j^q X_k^r X_\ell^s (g_{ps}g_{qr} - g_{pr}g_{qs}) \\ &= K(\delta_{i\ell}\delta_{jk} - \delta_{ik}\delta_{j\ell}). \end{aligned}$$

or

$$\Theta_{ij} = -K\vartheta_i \wedge \vartheta_j. \tag{38}$$

Since this conclusion goes backwards, we see that condition (38) charaterizes a Riemann metric to be isotropic, with the common sectional curvature K.

Theorem 8.2.4 (F. Schur). *A Riemann manifold of dimension ≥ 3, which is isotropic at every point, has a constant sectional curvature for all plane elements.*

To prove the theorem we substitute the expressions (38) into the Bianchi identities (13), getting

$$dK \wedge \vartheta_i \wedge \vartheta_j = 0.$$

It follows that dK is a linear combination of ϑ_i, ϑ_j for any two i, j, $i \neq j$. When $n \geq 3$, this is possible only when $dK = 0$, which means that K is an absolute constant.

Definition 8.2.5. A Riemann manifold is said to be of constant curvature, if the sectional curvature is the same for all plane elements.

Theorem 8.2.6 (Synge). *Let g be a geodesic and S a surface through g. At any point O of g the Gaussian curvature of S is $\leq$ the sectional curvature in the tangent plane of S. Moreover, they are equal at all points of g if and only if the tangent planes of S are parallel along g.*

To prove this theorem we consider the family F of rectangular frames introduced in the beginning of this section. Denote by E the tangent plane to S at O and by K the Gaussian curvature of S at O. From (35) we get

$$K = R(O,E) + \sum_r (A_{r11}A_{r22} - A_{r12}^2)_0, \tag{39}$$

where the subscript 0 denotes the value of the quantity in question at O. In the family F of frames we consider the sub-family F' consisting of frames $pe_1 \cdots e_n$ satisfying the conditions: 1) p lies on g; 2) e_1 is tangent to g at p. We denote by $i : F' \to F$ the identity mapping. Then we have $i^*\vartheta_2 = 0$, and $i^*\vartheta_1 \neq 0$. When the family of frames along g is developed into on Euclidean space, the image frames $p'e_1' \cdots e_n'$ satisfy the differential system

$$dp' = (i * \vartheta_1)e_1', \qquad de_1' = (i * \vartheta_{12})e_2' + \sum_r (i * \vartheta_{1r})e_r'.$$

Since g is a geodesic, we have

$$i * \vartheta_{12} = 0, \qquad i * \vartheta_{1r} = 0.$$

This implies that

$$A_{r11} = 0.$$

Therefore, we have

$$K = R(O,E) - \sum_r (A_{r12}^2)_0.$$

which implies the first statement of the theorem.

From the last equation we also see that $K = R(O,E)$ if and only if $(A_{r12})_0 = 0$. The last condition is equivalent to

$$i^*\vartheta_{2r} = 0.$$

Since $i^*\vartheta_{12} = 0$, we get $de_2' = 0$. But this is equivalent to saying that the vectors e_2' are parallel along g. Hence the second statement of the theorem is proved.

8.3 Normal coordinates; Existence of convex neighbourhoods

Normal coordinates in a Riemann manifold are special cases of affine normal coordinates treated in §7.4. We shall only discuss those aspects which are peculiar to the Riemannian case. In particular, it will be proved that every point O has a convex neighbourhood, that is, a neighbourhood U such that any two of its points can be joined by exactly one geodesic belonging to U. A similar theorem on affine connections is valid, when the geodesics are replaced by paths, but our proof does not apply to the affine case.

We recall a few results developed in the affine case. Let O be a point of M. We take a rectangular frame F_0 at O and displace it parallelly along the geodesics through O. Since the frames obtained will all be rectangular, we can lower superscripts and use subscripts throughout. The normal coordinates are defined by

$$x_i = \alpha_i s, \qquad \alpha_1^2 + \cdots + \alpha_n^2 = 1, \tag{40}$$

where α_i are the components of the unit tangent vector of a geodesic through O relative to F_0 and s is the arc length along the geodesic. (Notice that we have changed our notation in writing x_i as the normal coordinates). If ϑ_i, ϑ_{ij} restrictions of the same differential forms to the family of frames defined above, we can write

$$\begin{aligned} \vartheta_i &= \alpha_i ds + \overline{\vartheta}_i, \\ \vartheta_{ij} &= \overline{\vartheta}_{ij}, \end{aligned} \tag{41}$$

where $\overline{\vartheta}_i$, $\overline{\vartheta}_{ij}$ do not involve ds. Moreover, they are solutions of the system of partial differential equations

$$\begin{aligned} \frac{\partial \overline{\vartheta}_i}{\partial s} &= d\alpha_i + \sum_j \alpha_j \overline{\vartheta}_{ji}, \\ \frac{\partial \overline{\vartheta}_{ij}}{\partial s} &= \sum_{k,\ell} S_{ijk\ell} \alpha_k \overline{\vartheta}_\ell, \end{aligned} \tag{42}$$

with the initial conditions

$$(\overline{\vartheta}_i)_0 = 0, \quad (\overline{\vartheta}_{ij})_0 = 0. \tag{43}$$

Writing

$$\overline{\vartheta}_i = s d\alpha_i + \sum_j A_{ij} d\alpha_j, \tag{44}$$

we see that the functions A_{ij} satisfy the initial conditions

$$(A_{ij})_0 = 0, \quad \left(\frac{\partial A_{ij}}{\partial s}\right)_0 = 0. \tag{45}$$

These imply in particular that the functions A_{ij}/s^2 are regular at 0.

Let U be a neighbourhood of O in which the normal coordinate system x_i is valid. The element of arc is equal to

$$\sum_i \vartheta_i^2 = ds^2 + 2ds.\sum_i \alpha_i \overline{\vartheta}_i + \sum \overline{\vartheta}_i^2.$$

By (42) we have

$$\frac{\partial}{\partial s}\left(\sum_i \alpha_i \overline{\vartheta}_i\right) = \sum_i \alpha_i \left(d\alpha_i + \sum_j \alpha_j \overline{\vartheta}_{ji}\right) = 0,$$

so that $\sum_i \alpha_i \overline{\vartheta}_i$ is equal to its value at $s = 0$, i.e.,

$$\sum_i \alpha_i \overline{\vartheta}_i = 0.$$

Hence the element of arc is equal to

$$\sum_i \vartheta_i^2 = ds^2 + \sum_i \overline{\vartheta}_i^2. \tag{46}$$

From (46) it follows that the hypersurfaces $s =$ const. and the radial geodesics $\alpha_i =$ consts. are orthogonal to each other.

We wish to prove that for every point $p \in U$ the shortest arc between O and p is geodesic g joining them, and that this is the only shortest arc. In fact, let C be any parametrized curve joining O, p. For simplicity we suppose it to be defined by $\alpha_i = \sum_i(s)$. If s_0 is the length of g from O to p, we have

$$s_0 = \int_0^{s_0} ds \leq \int_0^{s_0} \sqrt{ds^2 + \sum_i \overline{\vartheta}_i^2},$$

where the last integral is taken along C and is equal to the length of C from O to p. This proves that g is the shortest arc joining O, p. Equality sign holds in the above relation, only if $\overline{\vartheta}_i = 0$ along C. By (44), we get

$$d\alpha_i + \sum_j \frac{A_{ji}}{s} d\alpha_j = 0.$$

Since $A_{ji}/s \to 0$ as $s \to 0$, we have $d\alpha_i = 0$ along C, U is sufficiently small. But the latter means that C is a geodesic.

For latter reference we state our results in the following lemma (Cf. Lemma 7.4.1).

Lemma 8.3.1. *Every point O of a Riemann manifold M has a neighbourhood U in which a normal coordinate system is valid and which has the properties: 1) Every point $p \in U$ has a neighbourhood covered by a normal coordinate system at p and containing U; 2) The geodesic joining O and p is the only curve of shortest length in M joining these two points.*

In the study of the geometry on a Riemann manifold, one of the effective tools is to introduce a suitable metric in the space and apply results developed in the topology of metric spaces. That such a natural metric exists is asserted by the following theorem.

Theorem 8.3.2. *Let M be a connected Riemann manifold. For any two points $p,q \in M$ we define the distance (p,q) to be $\inf \widehat{pq}$, where $\widehat{pq}$ is the arc length of any rectifiable arc joining p,q. This definition of distance makes M into a metric space, which induces the same topology as the given topology in M. If there is an arc g joining p, q whose length is equal to (p,q), g must be a geodesic. Such a geodesic arc has the property:*

$$(r_1,r_2)+(r_2,r_3)=(r_1,r_3), \tag{47}$$

for any three points r_1, r_2, r_3 of g, with r_2 between r_1 and r_3. Conversely, an arc satisfying the condition (47) *for any three points is a geodesic and has the property that the distance between any two of its points its equal to the arc length.*

Geodesic arcs with the property (47) will be called minimizing. It is to be remarked that not any two points can be joined by a minimizing geodesic arc. Nor is it true that every geodesic arc is minimizing. Examples showing this can be readily constructed on the ordinary sphere with a point removed.

We first prove that there is always a rectifiable arc joining p, q. In fact, from the connectedness of M it follows that there is a sequence of coordinate neighbourhoods $U_1,\dots,U_m$, such that $p \in U_1$, $q \in U_m$, $U_i \cap U_{i+1} \neq 0$, $i=1,\dots,m-1$. We take a point $p_i \in U_i \cap U_{i+1}$. Then any two consecutive points of the sequence $p, p_1, \dots, p_{m-1}$, q lie in the same coordinate neighbourhood and can be joined by a rectifiable arc. These arcs make up a rectifiable arc joining p, q.

This distance function clearly has the properties:

$$\begin{aligned} &(p,p)=0, \quad (p,q)=(q,p), \\ &(p,r) \leq (p,q)+(q,r). \end{aligned} \tag{48}$$

By Lemma 8.3.1, we get $(p,q)>0$, if $p \neq q$. Using normal coordinates, we also see that, if p_i is a sequence of points, $(p,p_i) \to 0$ if and only if $p_i \to p$, the latter being understood in the sense of the given topology of M. Therefore, our metrisation of M does not change its topology.

Suppose now g to be an arc joining p, q with length (p,q). The property of g to be a geodesic is a local property. If, therefore, r is a point of g at which g is not a geodesic, we can, by Lemma 8.3.1, replace g by a shorter arc joining p, q, contradicting our hypothesis. Hence g is a geodesic. Let r_1, r_2, r_3 be three points of g, with r_2 between r_1 and r_3. Then we have

$$\widehat{pr_1}+\widehat{r_1r_2}+\widehat{r_2r_3}+\widehat{r_3q}=\widehat{pq}=(p,q),$$

from which it follows that

$$(p,r_1)+(r_1,r_2)+(r_2,r_3)+(r_3,q) \le (p,q).$$

The triangle inequality shows that the equality sign must hold and that

$$(r_1,r_2) = \widehat{r_1r_2}, (r_2,r_3) = \widehat{r_2r_3}.$$

A similar argument gives

$$(r_1,r_3) = \widehat{r_1r_3},$$

whence (47) follows.

Take a point $r_2 \in g$; let V be a neighbourhood of r_2 with the properties of Lemma 8.3.1. For definiteness suppose r_2 to be an inner point of g. Let $r_1, r_3 \in g \cap V$, such that r_2 is between r_1 and r_3. Since the minimum arc length between r_1, r_3 is realized only on a geodesic, condition (47) is possible only when r_2 lies on the geodesic joining r_1, r_3. It follows that g is a geodesic at r_2. Similar arguments prove that g is also a geodesic at its end-points.

Let p, q be any two points of g. g being a geodesic, every point has a neighbourhood in which its distance to a point of g is realized by the arc length on g. Since g is compact, it can be covered by a finite number of these neighbourhoods. It follows that there is a finite number of points $r_0 = p, r_1, \dots, r_m = q$, such that $(r_i, r_{i+1}) = \widehat{r_ir_{i+1}}$, $i = 0, 1, \dots, m-1$. Then we have

$$(p,q) = (r_0,r_m) = \sum_{i=0}^{m-1}(r_i,r_{i+1}) = \sum_{i=0}^{m-1}\widehat{r_ir_{i+1}} = \widehat{pq}$$

which shows that (p,q) is equal to the arc length $\widehat{pq}$ on g.

Lemma 8.3.3. *At every point $O \in M$ there is a number $c > 0$ such that every hypersphere Σ of radius $< c$ about O has the preperties: 1) Every point of Σ can be joined to O by a unique geodesic of has the shortest length; 2) Every geodesic tangent to Σ lies strictly in the outside of Σ (in a neighbourhood of the point of contact).*

Let x_i be a normal coordinate system at O, and let γ_{ijk} be the components of the connection of Levi-Civita relatively to x_i. The differential equations of the geodesics:

$$\frac{d^2x_i}{ds^2} + \sum_{j,k}\gamma_{ijk}\frac{dx_j}{ds}\frac{dx_k}{ds} = 0.$$

are satisfied by the equations

$$x_i = \alpha_i s, \quad \alpha_i = \text{const.},$$

so that we have

$$\sum_{j,k}\gamma_{ijk}\alpha_j\alpha_k = 0. \tag{49}$$

Since the values of α_i are arbitrary, we must have

$$(\gamma_{ijk})_0 = 0. \tag{50}$$

Let

$$2F(x_1,\ldots,x_n) \equiv x_1^2 + \cdots + x_n^2 - \rho^2 = 0 \tag{51}$$

be the equation of Σ, whose radius is $\rho > 0$. Let

$$x_i = x_i(\alpha) \tag{52}$$

be the equations of a geodesic tangent to Σ at a point p with σ as its arc length measured from p. Then we have

$$F(x_i(0)) = 0, \quad \sum_i \left(x_i \frac{dx_i}{d\sigma}\right)_{\sigma=0} = 0.$$

But we find

$$\frac{d}{d\sigma}F(x_i(\sigma)) = \sum_i x_i \frac{dx_i}{d\sigma},$$

$$\frac{d^2}{d\sigma^2}F(x_i(\sigma)) = \sum_{i,j}\left(\delta_{ij} - \sum_k x_k \gamma_{kij}\right)\frac{dx_i}{d\sigma}\frac{dx_j}{d\sigma}.$$

It follows that

$$F(x_i(\sigma)) = \frac{1}{2}\sum_{i,j}\left(\delta_{ij} - \sum_k x_k \gamma_{kij}\right)_{\sigma=0}\left(\frac{dx_i}{d\sigma}\right)_{\sigma=0}\left(\frac{dx_j}{d\sigma}\right)_{\sigma=0}\sigma^2 + \cdots \tag{53}$$

In a sufficiently small neighbourhood of O, γ_{kij} can be made arbitrarily small. Hence the coefficient of σ^2 in the above expression is strictly positive. This proves that the geodesic (52) lies strictly outside of Σ in a neighbourhood of p, and that its only common point with Σ is p itself.

Theorem 8.3.4. *Every point O of a Riemann manifold has a convex neighbourhood consisting of all points p with $(0,p) < \varepsilon$ for sufficiently small ε such that any two of its points can be joined by exactly one geodesic lying in the neighbourhood.*

Let $c > 0$ be a number with the properties of Lemma 8.3.3. Let V be a neighbourhood satisfying the conditions of Lemma 8.3.1, and W_0 an ε-neighbourhood contained in V. We choose η satisfying the inequalities

$$\eta \le \frac{1}{4}c, \quad \eta \le \frac{1}{4}\varepsilon,$$

and denote by W an η-neighbourhood of O. We shall prove that W is convex.

In fact, let $q, q' \in W$. Since they belong to V, there is a unique minimizing geodesic β joining q, q'. Let m be any point on β. We have

$$(q,q') \leq (0,q) + (0,q') \leq 2\eta \leq \frac{1}{2}\varepsilon,$$

and, since β is a minimizing geodesic.

$$(0,m) \leq (m,q) + (q,0) \leq (q',q) + (q,0) < \frac{3}{4}\varepsilon.$$

It follows that $m \in W_0$ or $\beta \subset W_0$.

Suppose that β does not belong to W. Let $m \in \beta$ be such that the distance $(0,m)$ is a maximum. Then $(0,m) \geq \eta$. Since $(0,q) < \eta$, $(0,q') < \eta$, m is an interior point of the arc β. How we have

$$\begin{aligned}(0,m) &\leq (0,q) + (q,m) < (0,q) + (q,q') \\ &\leq 2(0,q) + (0,q') \\ &\leq 3\eta < c.\end{aligned}$$

By Lemma 8.3.3 there is a point $m' \in \beta$ in every neighbourhood of m such that $(0,m') > (0,m)$. But this contradicts the definition of m. It follows that $\beta \subset W$, and the theorem is proved.

8.4 Gauss-Bonnet formula

Cf. Chern, Topics in differential geometry, Princeton 1951, pp. 38–41.

8.5 Completeness

Completeness is a property of a Riemann manifold to ensure that it is not a proper open submanifold of another. As we shall show by examples, it is stronger than the latter property. However, it seems to be an appropriate notion, because we have, among other things, the theorem: Any two points on a complete Riemann manifold can be joined by a minimizing geodesic. The proof of this theorem will be the main purpose of this section.

In the following discussions we shall make use of the metric in M defined in §8.3. It will be convenient to introduce some definitions and prove some theorems pertaining to arbitrary metric spaces.

Definition 8.5.1. An arc in a metric space m is a continuous image of a closed interval $I : 0 \leq t \leq 1$. A path is a continuous image of an interval $0 \leq t < 1$, open at one end. An arc $p(t)$ is called a segment if

$$(p(t_1),p(t_2))+(p(t_2),p(t_3))=(p(t_1),p(t_3)), \tag{54}$$

whenever $0\le t_1\le t_2\le t_3\le 1$. In these definitions the unit interval can be replaced by any interval, respectively closed or open at one end. It follows that the restriction of an arc to image of a sub-interval is an arc, called a subarc. Clearly, a subarc of a segment is a segment. A path is called a ray, if, 1) It is a closed subset of m; 2) Each of its subarcs is a segment. If the distance $(p(0),p(t))$ of a ray $p(t)$ is bounded, its least upper bound is called the length of the ray.

The main theorem of this section is the following.

Theorem 8.5.2. *To any two points $p,q\in M$ there is either a segment joining them or a ray with origin p whose points x satisfy the relation $(p,x)+(x,q)=(p,q)$.*

We begin by proving some lemmas.

Lemma 8.5.3. *Let $a_1,\dots,a_m$ be points in a metric space m satisfying the relation.*

$$(a_1a_2)+\cdots+(a_{m-1}a_m)=(a_1a_m). \tag{55}$$

Then for any set of integers $i_1,\dots,i_k$, with $1\le i_1\le\cdots\le i_k\le m$, we have

$$(a_{i_1}a_{i_2})+\cdots+(a_{i_{k-1}}a_{i_k})=(a_{i_1}a_{i_k}). \tag{56}$$

Suppose that the relation (56) is not true for a certain set of integers $i_1,\dots,i_k$. By the triangle inequality we must have

$$(a_{i_1}a_{i_2})+\cdots+(a_{i_{k-1}}a_{i_k})>(a_{i_1}a_{i_k}).$$

It follows that

$$\begin{aligned}&(a_1a_2)+\cdots+(a_{m-1}a_m)\\ &\ge(a_1a_{i_1})+(a_{i_1}a_{i_2})+\cdots+(a_{i_{k-1}}a_{i_k})+(a_{i_k}a_m)\\ &>(a_1a_{i_1})+(a_{i_1}a_{i_k})+(a_{i_k}a_m)\ge(a_1a_m),\end{aligned}$$

which contradicts (55).

Lemma 8.5.4. *Let a_ka_{k+1}, $k=1,\dots,m-1$, be segments in a metric space m satisfying the relation*

$$(a_1a_2)+\cdots+(a_{m-1}a_m)=(a_1a_m). \tag{55}$$

Then the arc $\gamma=\sum_{k=1}^{m-1}a_ka_{k+1}$ is a segment.

Consider first the case $m=3$. Obviously $a_1a_2+a_2a_3$ is an arc. To show that it is a segment we have to prove that, for any three of its points x,y,z with y between x,z we have $(xy)+(yz)=(xz)$. If all of them belong to a_1a_2 or to a_2a_3, this follows from- the hypothesis. Suppose that $x,y\in a_1a_2$ and $z\in a_2a_3$. Since a_1a_2 and a_2a_3 are segments, we have

$$(a_1x) + (xy) + (ya_2) = (a_1a_2),$$
$$(a_2z) + (za_3) = (a_2a_3).$$

From (55) it follows that

$$(a_1x) + (xy) + (ya_2) + (a_2z) + (za_3) = (a_1a_2) + (a_2a_3) = (a_1a_3).$$

Hence, by Lemma 8.5.3, we have

$$(xy) + (yz) = (xz).$$

The case $x \in a_1a_2$, $y, z \in a_2a_3$ can be treated similarly, and the general case follows from an obvious induction on m, with use made of Lemma 8.5.3.

Lemma 8.5.5. *Let β be a path $p(t)$, $0 \le t < 1$, in a metric space m, of which every subarc is a segment. Then either β is closed in m and is therefore a ray; or* lim *$p(t)$ exists, and then, if p is the limit, $\beta + p$ is a segment.*

Suppose that β is not closed. Let p be a limit paint of β not belonging to β. There exists a sequence $t_\nu \to 1$, $0 \le t_\nu < 1$, such that $a_\nu = p(t_\nu) \to p$. Let t'_μ, $0 \le t'_\mu < 1$, be any sequence tending to 1. We shall prove that $p(t'_\mu) \to p$.

To any t'_μ, with μ sufficiently large, let $t_{\nu(\mu)} < t'_\mu$ be such there is no t_ν in the interval $t_{\nu(\mu)} < t < t'_\mu$. The set $\{t_{\nu(\mu)}\}$ is clearly not finite. Being a subsequence of a convergent sequence, the sequence $t_{\nu(\mu)} \to 1$ and $p(t_{\nu(\mu)}) \to p$. We put $t_{\mu_1} = t_{\nu(\mu)}$. On the other hand, let t_{μ_2} be such that $t'_\mu < t_{\mu_2}$, $\mu_1 < \mu_2$. Then the sequence a_{μ_2} also converges to p, as $\mu \to \infty$. Since $a_{\mu_1}a_{\mu_2}$ is a segment, we have

$$(a_{\mu_1}p(t'_\mu)) + (p(t'_\mu)a_{\mu_2}) = (a_{\mu_1}a_{\mu_2}),$$

whence

$$(a_{\mu_1}p(t'_\mu)) \le (a_{\mu_1}a_{\mu_2}).$$

It follows that

$$\begin{aligned}(pp(t'_\mu)) &\le (pa_{\mu_1}) + (a_{\mu_1}p(t'_\mu))\\ &\le (pa_{\mu_1}) + (a_{\mu_1}a_{\mu_2}) \le 2(pa_{\mu_1}) + (pa_{\mu_2}).\end{aligned}$$

As $\mu \to \infty$, each of the summands on the right-hand side tends to O. Therefore $p(t'_\mu) \to p$.

It follows that $p(t)$, $0 \le t \le 1$, is an arc, if we define $p(1) = p$. It will be a segment, if, for any three of its points x, y, z, we have $(xy) + (yz) = (xz)$. This follows from hypothesis, unless $y \ne p$, $z = p$. In the latter case, we use, for sufficiently large ν, the relation

$$(xy) + (ya_\nu) = (xa_\nu).$$

From this the desired relation follows, as $\nu \to \infty$.

The above lemmas are valid in any metric space. In order for Theorem 8.5.2 to be true we need a local property, as given by the following lemma.

Lemma 8.5.6. *To every point p of a Riemann manifold there exists a number $\rho(p) > 0$, with the following properties: 1) If q is a point with $(pq) \leq \rho(p)$, there exists a segment pq; 2) If q is a point with $(pq) > \rho(p)$, there is a point $x \in M$, with $(px) = \rho(p)$, and $(px) + (xq) = (pq)$.*

We take $\rho = \rho(p) > 0$ to be less than the radius of a neighbourhood of p, in which a normal coordinate system at p is valid. Property 1) is clearly fulfilled. Let q be a point with $(pq) > \rho$, and let β_ν be a sequence of rectifiable arcs joining p, q, whose lengths s_ν tend to the distance (pq). Let x_ν be the point on β_ν with $(px_\nu) = \rho$. The points x_ν lie on the sphere of radius ρ about p. Since the sphere is compact, there is a subsequence of x_ν tending to a limit point x on the sphere. For simplicity of notation we denote this subsequence still by x_ν. Then we have $(x_\nu q) \leq s_\nu - \rho$, so that

$$(pq) \leq (px_\nu) + (x_\nu q) \leq s_\nu.$$

As $\nu \to \infty$, we have therefore $(px_\nu) + (x_\nu q) \to (pq)$ or $(px) + (xq) = (pq)$, as was to be proved.

We can now proceed to give a proof of Theorem 8.5.2. Suppose that there is no segment joining the points p, q. We put $a_1 = p$, and shall construct a sequence of segments $a_i a_{i+1}$, $i = 1, 2, \ldots$, satisfying the condition

$$(a_1a_2) + \cdots + (a_{k-1}a_k) + (a_kq) = (a_1q). \tag{C_k}$$

This will, of course, be done by induction on k. We notice first that, by Lemma 8.5.3, (C_k) implies

$$(a_1a_2) + \cdots + (a_{k-1}a_k) = (a_1a_k), \tag{A_k}$$

$$(a_1a_k) + (a_kq) = (a_1q). \tag{B_k}$$

Also, by Lemma 8.5.4, there is no segment joining a_k and q.

Suppose that segments $a_1a_2, \cdots, a_{k-1}a_k$ have been constructed so as to satisfy (C_k), and let us construct a_{k+1}. We consider all $x \in M$, such that: 1) $(a_kx) + (xq) = (a_kq)$; 2) A segment a_kx exists. By Lemma 8.5.6, the set of these x is not empty. Put $T_{k+1} = \limsup(a_kx)$. Then we have

$$0 < T_{k+1} \leq (a_kq).$$

We choose a_{k+1} satisfying the conditions: 1) There is a segment a_ka_{k+1}; 2) $(a_ka_{k+1}) + (a_{k+1}q) = (a_kq)$; 3) $(a_ka_{k+1}) \geq \frac{1}{2}T_{k+1}$. Then we have, by condition 2) and (C_k),

$$\begin{aligned}&(a_1a_2) + \cdots + (a_{k-1}a_k) + (a_ka_{k+1}) + (a_{k+1}q)\\ =&(a_1a_2) + \cdots + (a_{k-1}a_k) + (a_kq) = (a_1q),\end{aligned}$$

which is precisely (C_{k+1}). Hence the induction is complete.

The set

$$\beta = \sum_{i=1}^{\infty} a_i a_{i+1}$$

can be considered as a path in an obvious way. By Lemma 8.5.4, every subarc $a_1 a$ is a segment, so that the path β satisfies the hypotheses of Lemma 8.5.5. We shall prove that the sequence a_k has no limit point, so that β is a ray. For this purpose assume that $a_k \longrightarrow m$. Then $\beta + m$ is a segment. As $k \longrightarrow \infty$, B_k) gives

$$(a_1 m) + (mq) = (a_1 q).$$

It follows that

$$(a_1 a_k) + (a_k m) + (mq) = (a_1 m) + (mq) = (a_1 q).$$

Since no segment $a_1 q$ exists, $m \neq q$. By Lemma 8.5.6 there is an $x \neq m$ such that $(mx) + (xq) = (mq)$ and that a segment mx exists. For this x we have

$$(a_k m) + (mx) + (xq) = (a_k m) + (mq) = (a_k q),$$

which gives

$$(a_k m) + (mx) = (a_k x).$$

Hence $a_k m + mx$ is a segment. From our definition of T_{k+1} we have

$$T_{k+1} \geqq (a_k x) \geq (mx) > 0, \qquad \text{for all } k \geqq 1.$$

$$\frac{1}{2} \sum_{k=1}^{\infty} T_{k+1} \leq \sum_{k=1}^{\infty} (a_k a_{k+1}) \leq (a_1 q),$$

which is possible only when $T_{k+1} \longrightarrow 0$ as $k \longrightarrow \infty$. But these two facts contradict with each other. This proves that β is a ray.

It remains to show that $(a_1 y) + (yq) = (a_1 q)$ for all $y \in \beta$. As a matter of fact, for sufficiently large k, we have

$$(a_1 y) + (y a_k) = (a_1 a_k),$$

whence

$$(a_1 y) + (y a_k) + (a_k q) = (a_1 a_k) + (a_k q) = (a_1 q).$$

This gives the desired relation, by Lemma 8.5.3. Hence the proof of Theorem 8.5.2 is complete.

Definition 8.5.7. A connected Riemann manifold is called complete, if it is a complete metric space in terms of the metric of Theorem 8.3.4.

Theorem 8.5.8. *Any two points of a complete Riemann manifold can be joined by a minimizing geodesic (or a segment).*

This is a consequence of Theorem 8.5.1 and the following theorem.

Theorem 8.5.9. *A complete Riemann manifold can be characterized by one of the following properties:*

1) There is no ray of finite length;
2) Every infinite bounded set has a limit point.

We first prove that completeness implies 1). Suppose, in fact, that $p(t)$, $0 \leq t < 1$, is a ray of finite length L. We take $t_k = 1 - 1/2^k$, and put $a_k = p(t_{k-1})$, $k = 1, 2, \ldots$. Then we have

$$\sum_{k=1}^{\infty}(a_k a_{k+1}) \leq L.$$

It follows that the sequence a_k is a fundamental sequence. But it has no limit point, which implies that M is not complete.

Next we prove that 1) implies 2). By Theorem 8.5.2, any two points of M can be joined by a segment. Let S be a bounded infinite set of M, so that there exists a point $a \in M$ and a constant K such that $(a,x) < K$ for all $x \in S$. The points a and x can be joined by a segment or a minimizing geodesic. There exists, therefore, a sequence of distinct points $x_i \in S$ such that the length s_i of the geodesic ax_i tends to $s \leq K$ and that the tangent vector to ax_i at a tends to a vector v through a. Let y be the point on the geodesic through a tangent to v with $ay = s$. From the theorem that the solution of the differential equations of the geodesics depends continuously on the initial conditions it follows that the sequence x_i has y as limit point.

That 2) implies the completeness of M is a well-known theorem in topology. Thus the theorem is completely proved.

Corollary 8.5.10. *A compact Riemann manifold is complete.*

In the discussion of the concept of completeness the problem which faces us is to impose a restriction on M so that it will not be a proper open submanifold of another Riemann manifold. The notion of completeness achieves this purpose. In fact, suppose M be a proper open submanifold of a Riemann manifold M'. Let $p\overline{\in}M$ be a boundary point of M. Let $q \in M$ be a point in a sufficiently small neighbourhood of p. Clearly the segment joining q to p contains a finite ray of M. This proves that M is not complete.

On the other hand, completeness is a stronger restriction than the above property, which may be called *non-continuability*. To show this, the simplest example is given by the universal covering manifold of $E - O$, where E is the Euclidean plane and O a point of E. This manifold is homeomorphic to the Euclidean plane. It can be proved that it is non-continuable, while there are pairs of points which can not be joined by minimizing geodesics. This example also shows that non-continuability of a Riemann manifold is not sufficient for Theorem 8.5.8 to be true. For this reason and for at least another reason to be given in the next section in the discussion of manifolds of constant curvature it seems that completeness is an appropriate restriction to be imposed on a Riemann manifold.

By passing to the universal covering manifold of M, we can derive from Theorem 8.5.8 the following theorem.

Theorem 8.5.11. *In every free homotopy class $\neq 0$ of closed curves of a compact Riemann manifold there is a closed geodesic, which is the shortest closed curve in the class.*

We shall not give the proof of this theorem here, although it is not difficult.

8.6 Manifolds of constant curvature

Among Riemann manifolds those of constant curvature are perhaps the simplest and geometrically the most important. We begin by exhibiting such manifolds. It should be remarked that two Riemann manifolds with the same constant curvature are locally equivalent.

We try to determine the function ρ such that the element of arc given by

$$ds^2 = \frac{1}{\rho^2}(du_1^2 + \cdots + du_n^2),$$

where

$$\rho = \rho(r), r^2 = u_1^2 + \cdots + u_n^2,$$

is of constant curvature. Consider at each point the rectangular frame whose ith vector is the unit vector tangent to the u_i-curve. Denote by ϑ_i, ϑ_{ij} the restrictions to this family of the same differential forms in the bundle of rectangular frames. We have

$$\vartheta_i = \frac{1}{\rho} du_i.$$

It follows that

$$\begin{aligned} d\vartheta_i &= -\frac{1}{\rho_2} \sum_j \frac{\partial \rho}{\partial u_j} du_j \wedge du_i \\ &= \sum_j \vartheta_{ij} \wedge \vartheta_j = \frac{1}{\rho} \sum_j \vartheta_{ij} \wedge du_j, \end{aligned}$$

or

$$\sum_j \left(\vartheta_{ij} - \frac{\partial \log \rho}{\partial u_j} du_j + \frac{\partial \log \rho}{\partial u_i du_j} \right) \wedge du_j = 0.$$

From this we get

$$\vartheta_{ij} = \frac{\partial \sigma}{\partial u_j} du_i - \frac{\partial \sigma}{\partial u_i} du_j = \frac{\sigma'}{r}(u_j du_i - u_i du_j),$$

where the dash denotes differentiation with respect to r, and where

$$\sigma = \log \rho .$$

The Riemann-Christoffel tensor is then given by the formula

$$\begin{aligned} d\vartheta_{ij} - \sum_k \vartheta_{ik} \wedge \vartheta_{kj} \\ = \left\{ \frac{1}{r}\left(\frac{\sigma'}{r}\right)^1 + \left(\frac{\sigma'}{r}\right)^2 \right\} \left(\sum_k u_k du_k \right) \wedge (u_j du_i - u_i du_j) \\ + \left(-2\frac{\sigma'}{r} + \sigma'^2 \right) du_i \wedge du_j . \end{aligned}$$

In order that the Riemann metric be of constant curvature it is necessary that

$$\frac{1}{r}\left(\frac{\sigma'}{r}\right)^1 + \left(\frac{\sigma'}{r}\right)^2 = 0.$$

This differential equation can be integrated, the general integral being

$$\sigma = \log(r^2 - 2A)B,$$

where A, B are constants. The corresponding function ρ is then

$$\rho = B(r^2 - 2A)$$

while the curvature is given by

$$K = \rho^2 \left(2\frac{\sigma'}{r} - \sigma'^2 \right) = -8AB^2 .$$

For simplicity we choose the constants A, B to satisfy the relation $-2AB = 1$. Then $K = 4B$ and

$$\rho = 1 + \frac{K}{4} r^2 .$$

It follows that the element of arc given by

$$ds^2 = \frac{du_1^2 + \cdots + du_n^2}{\left\{ 1 + \frac{K}{4}(u_1^2 + \cdots + u_n^2) \right\}} \tag{57}$$

is of constant curvature K.

The above considerations are local. We shall investigate the domain in the space of u_i, in which the above element of arc is valid. If $K = 0$, this extends to the whole space, which is the Euclidean space. If $K < 0$, the metric is defined in the interior of a hypersphere:

$$1 + \frac{K}{4}(u_1^2 + \cdots + u_n^2) > 0.$$

As a point tends to the boundary of the hypersphere, $ds^2 \to \infty$. This proves that the Riemann manifold is complete. It is called the *hyperbolic space* of n dimensions.

If $K > 0$, the metric is again defined in the whole space. But the Riemann manifold so defined is not complete. In fact, by stereographic projection, it can be mapped homeomorphically onto a hypersphere of radius $\sqrt{K}$ in a Euclidean space of $n+1$ dimensions, with a point removed, such that its metric is mapped into that of the hypersphere. The hypersphere is a Riemann manifold of constant positive curvature, called the *spherical space*. Being compact, it is complete. The hyperbolic and spherical spaces are the spaces of non-Euclidean geometry. We notice that, for $n \geq 2$, both are simply connected. The same is of course true of the Euclidean space.

Definition 8.6.1. Two Riemann manifolds are called isometric, if there is a differentiable homeomorphism between them which maps one element of arc into the other. They are called locally isometric if such a homeomorphism exists between two neighbourhoods.

Theorem 8.6.2. *A simply connected complete Riemann manifold of constant curvature K is isometric to one of the spaces: 1) the Euclidean space, if $K = 0$; 2) the hiperbolic space, if $K < 0$; 3) the spherical space, if $K > 0$.*

We denote by M^* the Euclidean, hyperbolic, or spherical space, according to the sign of K. Take $O \in M$, $O^* \in M^*$, and take a frame each at O and O^*. For $p^* \in M^*$ let $g^* = O^*p^*$ be the geodesic joining O^*, p^*. g^* is uniquely determined, except when M^* is the spherical space and p^* is the antipodal point $\overline{O}^*$ of O^*, a case we exclude at the moment. Let g be the geodesic through O, whose unit tangent vector has the same components relative to the frame at O as those of the unit tangent vector to g^* relative to the frame at O^*. Let p be the point on g such that the arc length Op is equal to (O^*, p^*). We define a mapping $f : M^* \to M$ be putting $f(p^*) = p$. Since M is complete, f is onto.

The mapping f is an isometry at the points p^* belonging to a neighbourhood of O^*. (f is said to be an isometry at a point if it is such in a neighbourhood of that point). Moreover, the points M^* at which of f is an isometry form clearly an open set. Let $p^* \in M^*$ be a point such that f is an isometry at all points of the geodesic O^*p^* open at the end p^*. We shall prove that it is also an isometry at p^*. In fact, p being the image point of p^*, there is an ε-neighbourhood U of p which is isometric to an ε-neighbourhood U^* of p^*. Let $q^* \neq p^*$ be a point of the geodesic O^*p^*, and V^* a neighbourhood of q^* contained in U^*, such that $f|V^*$ is an isometry. Put $q = f(q^*)$, $V = f(V^*)$. We can determine an isometry $h : U^* \to U$, satisfying $h^{-1}|V = f^{-1}|V$. Then f and h are identical in the intersection of U^* with the cone of geodesics joining O^* to the points of V^*. This intersection contains a neighbourhood of p^*. Since h is by construction an isometry, the same is true of f at p^*.

We have to complete the definition of f for the case that M^* is the spherical space and $\overline{O}^*$ is the antipodal point of O^*. Denote by π the distance $(O^*, \overline{O}^*)$. Let

$p_1^*(\varepsilon)$, $p_2^*(\varepsilon)$ be points on two geodesics through O^*, at a distance $\pi - \varepsilon$ from O^*, ε being arbitrarily small, then $(p_1^*(\varepsilon), p_2^*(\varepsilon)) \to 0$ as $\varepsilon \to 0$. Since f is an isometry at all points $\neq \overline{O}^*$, we have

$$(fp_1^*(\varepsilon), fp_2^*(\varepsilon)) \to 0 \quad \text{as} \quad \varepsilon \to 0.$$

This proves that the points on the geodesics through O and at a distance π from O are identical. This point is then defined to be the image of $\overline{O}^*$ under f by continuity the mapping f is also an isometry at $\overline{O}^*$.

We shall now show that the mapping f defines M^* to be a covering manifold of M. For this purpose it is enough to prove: Every point $p \in M$ has a neighbourhood U such that $f^{-1}(U)$ is a union of disjoint open sets each mapped homeomorphically by f onto U. Suppose this not the case. Since f is locally an isometry, it would mean that to every $\varepsilon > 0$ there are points p_1^*, p_2^*, $p_1^* \neq p_2^*$, such that $f(p_1^*) = f(p_2^*) = p$ and $(p_1^*, p_2^*) < \varepsilon$. Let γ^* be the geodesic joining p_1^*, p_2^*. Then $f(\gamma^*)$ will be a closed geodesic through O, whose length is arbitrarily small. But this is impossible.

If follows that M^* is a covering manifold of M. Since M is itself simply connected, the continuous mapping f is a homeomorphism and is an isometry between M and M^* in the large.

Definition 8.6.3. A complete Riemann manifold of constant curvature is called a space form. It is called Euclidean hyperbolic, or spherical, according as the curvature is $= 0$, < 0, or > 0.

Corollary 8.6.4. *A Euclidean, hyperbolic, or spherical space form has as universal covering manifold respectively the Euclidean, hyperbolic, or spherical space.*

This corollary reduces the problem of the determination of the space forms to that of the determination of discontinuous groups of Euclidean and non-Euclidean motions, such that each of its motions has no fixed point. We recall that a group of transformations in the space M is called *discontinuous*, if, for every point $p \in M$, the set $T(p)$, where T runs over the transformations of the group, has no limit point. On the other hand, a topological group is called *discrete* if its underlying topological space is discrete. Of course, a discontinuous group is discrete. In our case, the Euclidean or non-Euclidean space can be regarded as a space of left cosets G/H, where G is the group of motions and H the group of isotropy leaving a point fixed, and where a motion in the space is defined by left multiplication of the cosets. In all three cases the group of isotropy H is compact, and the notions of discontinuous and discrete subgroups of motions are identical, in view of the following theorem.

Theorem 8.6.5. *Let the subgroup $\Gamma \subset G$ operate on the homogeneous space G/H as left multiplications. If Γ is discrete and H is compact, then Γ is discontinuous.*

Suppose that Γ is not discontinuous. There is then a sequence of mutually distinct $\gamma_i \in \Gamma$ such that the sequence of cosets $\gamma_i aH$ has the limit bH. We can assume

$b \neq e$, the unit element of G. Since Γ is discrete, we can choose neighbourhoods U, V of e, b respectively, such that $VUU^{-1}V^{-1}$ contains no element $\neq e$ of Γ. The condition that bH is the limit of the cosets $\gamma_i aH$ means that, for almost all i, there are $V_i \in V$, $b_i \in H$, satisfying

$$v_i^{-1}\gamma_i a = h_i.$$

Since H is compact, there is an $H \in H$ such that $h_i h^{-1} \in U$, for infinitely many I. In particular, there are $i \neq j$, with

$$h_i = u_i h, \qquad h_j = u_j h, \qquad u_i, u_j \in U.$$

Then we have

$$\gamma_i = v_i u_i h a^{-1}, \qquad \gamma_j = v_j u_j h a^{-1},$$

and

$$\gamma_i \gamma_j^{-1} = v_i u_i u_j^{-1} v_j^{-1} \in VUU^{-1}V^{-1}.$$

But $\gamma_i \gamma_k^{-1} \in \Gamma$, so that, $\gamma_i \gamma_j^{-1} = e$ or $\gamma_i = \gamma_j$. The last result contradicts the assumption that the γ_i are distinct. Hence the theorem is proved.

A space form is called *orientable* or *non-orientable*, according as its discrete group of covering transformations consists entirely of proper motions or not. It is important to remark that this group is isomorphic to the fundamental group of the manifold.

Perhaps the main interest in space form lies in the fact that they are a source of examples of topological manifolds. From the very definition such manifolds have rather special properties and are therefore useful in applications. We shall give in what follows a survey of this problem.

Some results of a general nature can be given for spherical forms, which constitute the simplest case. We take the spherical space as the Riemann manifold which is the unit hypersphere

$$x_1^2 + \cdots + x_{n+1}^2 = 1$$

in a Euclidean space of $n+1$ dimensions, with the metric induced by the Euclidean metric. A spherical motion is then given by an orthogonal matrix A of order $n+1$. Since the group of spherical motions is compact, every discrete subgroup is finite. It follows that every spherical form must behaves a finite fundamental group. Moreover, the characteristic roots of A are of absolute value 1. They cannot be $+1$, because the spherical motion in question has no fixed point. It follows that

$$\operatorname{sgn}|A| = \operatorname{sgn}|A - \lambda I|$$

for large λ. But the latter is $(-1)^{n+1}$. Since $|A| = \pm 1$, we have $|A| = (-1)^{n+1}$. If n is odd, $|A| = +1$. If n is even, every A is improper, except $A = 1$. Since the group of proper covering transformations is a subgroup of index 2 in Γ, Γ contains in this case only two elements, that is, I and $-I$. We summarize our results in the theorem.

Theorem 8.6.6. *Every spherical form has a finite fundamental group. If it is odd-dimensional, it is orientable. The only even-dimensional spherical forms are the sphere and the real projective space.*

The interest on spherical forms lies therefore in the odd-dimensional case. Its study is primarily a problem of group theory. Cf. G. Vincent, Comm. Math. Helv. 20, 117–171 (1947).

To study the Euclidean and hyperbolic forms we begin with the case $n = 2$. It can be proved that there are five two-dimensional Euclidean forms, as follows:

1) Euclidean plane;
2) Cylinder;
3) Torus;
4) Infinite Möbius band;
5)Klein bottle.

We shall prove that every closed orientable surface of genus $p > 1$ is a hyperbolic form. For this purpose we take in the hyperbolic plane a polygon P of $4p$ sides, which will be identified into a surface of genus p in a well-known fashion. The metric in the hyperbolic plane will give a metric on the resulting surface, if the sum of angles of the polygon P is 2π. Let α denote the sum of angles of P, and A its area. By the Gauss-Bonnet formula we get

$$\alpha = 2\pi(2p-1) + KA.$$

As $A \to 0, \alpha \to 2\pi(2p-1)$. On the other hand, P can expand in such a way that $\alpha \to 0$. By continuity it follows that there is a polygon P of $4p$ sides whose sum of interior angles is 2π. This proves our statement. Notice that the proof fails for $P = 1$. In fact, the sum of angles of any quadrangle in the hyperbolic plane is $< 2\pi$.

A Riemann metric on a non-orientable surface can be defined by taking the metric of constant curvature on its two sheeted covering surface. We summarize the results concerning the two-dimensional case in the theorem.

Theorem 8.6.7. *Every closed surface is a space form. The torus and Klein bottle are Euclidean forms, the sphere and projective plane are spherical forms, and all other surfaces are hyperbolic forms. Moreover, these cases are mutually exclusive.*

The last statement follows from the Gauss-Bonnet formula.

As mentioned above, there are numerous hyperbolic forms. However, in the higher dimensional case, it is not easy to construct compact ones. The known examples are the three-dimensional hyperbolic dodecahedral space and some examples due to F. Löbell. Cf. C. Weber und H. Sifort, Math. Z. 37, 237–253 (1933) and F. Lobell, Leipziger Berichte 83, 167–174 (1931).

The study of three-dimensional Euclidean forms is facilitated by the fact the so-called “Raumgruppen” are known. Cf. Hantsche-Wendt, Math. Annalen, 110, 593–611 (1934); W. Nowachi, Comm. Math. Helv. 1934.

Part IV Lecture Notes on Differentiable Geometry

9

Review of Surface Theory

9.1 Introduction

Let

$$x : M \to E^3 \tag{1}$$

be an immersed surface in the three dimensional euclidean space E^3. If u, v are local coordinates on M, it is given by the coordinate vector $x(u,v)$ as a smooth function of u, v, such that tangent vectors x_u, x_v are linearly independent. The unit normal vector $\nu(u,v)$ is then defined up to sign. Following Gauss, the local theory of surfaces is based on the first and second fundamental forms, which are

$$\begin{aligned} I &= (dx, dx) = Edu^2 + 2Fdudv + Gdv^2, \\ II &= -(dx, d\nu) = Ldu^2 + 2Mdudv + Ndv^2. \end{aligned} \tag{2}$$

The form I is positive define definite and the form II is defined up to a sign.

The radio

$$\frac{II}{I} = k_N\left(u, v, \frac{dv}{du}\right) \tag{3}$$

which depends on the point (u,v) and a tangent line T through it is called the normal curvature. Geometrically it is equal to the curvature at (u,v) of the curve of intersection of the plane spanned by the normal ν and T with the surface.

Let the point (u,v) be fixed. The critical values of k_N as a function of the tangent direction are called the principal curvatures. We will denote them by k_1 and k_2. Their elementary symmetric functions

$$\begin{aligned} H &= \frac{1}{2}(k_1 + k_2), \\ K &= k_1 k_2, \end{aligned} \tag{4}$$

are called, respectively, the mean curvature and the Gaussian curvature.

A Weingarten surface or a W-surface is one satisfying a relation

$$W(k_1,k_2)=0. \tag{5}$$

In particular, a surface with $H=0$ is called a minimal surface and a surface with $K=$ constant <0 is called a pseudospherical surface. In the following we will study such surfaces.

Remark 9.1.1. *If the surface is given in the non-parametric form*

$$z=z(x,y), \tag{6}$$

we have

$$H=\frac{(1+z_y^2)z_{xx}-2z_xz_yz_{xy}+(1+2z_x^2)z_{yy}}{(1+z_x^2+z_y^2)^{\frac{2}{3}}},$$

$$K=\frac{z_{xx}z_{yy}-z_{xy}^2}{(1+z_x^2+z_y^2)^2}.$$

In other words, equation (5) *is essentially a non-linear partial differential equation in two independent variables, expressed in a geometrical form.*

9.2 Moving frames

By a frame we mean a point x and an ordered set of vectors e_i, $i=1,2,3$, through x, such that

$$(e_i,e_j)=\delta_{ij}, \qquad (e_1,e_2,e_3)=1, \qquad i\le i,j\le 3. \tag{7}$$

Given such a frame, For instance the coordinate frame, there is exactly one orientation-preserving euclidean motion carrying it to another one. Hence the group G of all such motions can be identified with the space of all frames. We notice that its dimension is 6.

Since the vectors e_i are linearly independent, we can write

$$\begin{aligned} dx &= \sum_i \omega_i e_i, \\ de_i &= \sum_j \omega_{ij} e_j, \end{aligned} \tag{8}$$

where ω_i, ω_{ij} are linear differential forms, given by

$$\omega_i=(dx,e_i), \qquad \omega_{ij}=(de_i,e_j). \tag{9}$$

By differentiating the first equation of (7), we find

$$\omega_{ij}+\omega_{ji}=0 \tag{10}$$

In other words, the matrix

$$(\omega_{ij}) = \begin{bmatrix} \omega_{11} & \omega_{12} & \omega_{13} \\ \omega_{21} & \omega_{22} & \omega_{23} \\ \omega_{31} & \omega_{32} & \omega_{33} \end{bmatrix} \tag{10a}$$

is anti-symmetric. The forms ω_i, ω_{ij} are called the Maurer-Cartan forms of G.

We have

$$ddx = dde_i = 0,$$

which gives, by (8),

$$\begin{aligned} d\omega_i &= \sum_j \omega_{ij} \wedge \omega_j, \\ d\omega_{ij} &= \sum_k \omega_{ik} \wedge \omega_{kj}, \qquad 1 \leq i, j, k \leq 3. \end{aligned} \tag{11}$$

There are called the Maurer-Cartan equations. They play a fundamental role in differential geometry.

Suppose the surface (1) is oriented so that at every point $x \in M$ there is a well-defined unit normal vector. Consider the frames xe_i such that $x \in M$ and e_3 is the unit normal vector at x. Denote by E the space of surface frames. We have the map

$$\pi : E \longrightarrow M, \tag{12}$$

which sends the frame xe_i to the point xd. The inverse image $\pi^{-1}(x)$ consists of all frames which differ by a rotation

$$\begin{aligned} e_1' &= \cos \tau e_1 + \sin \tau e_2, \\ e_2' &= -\sin \tau e_1 + \cos \tau e_2, \end{aligned} \tag{13}$$

so that it is a circle with the angular coordinate τ. Since M is oriented, e_2 is completely determined by e_1 as the vector obtained from e_1 by rotating 90°, so that E can be considered as the space of all unit tangent vectors of M. E is called the unit tangent bundle of M and plays a fundamental role in surface theory. The sets $\pi^{-1}(x)$ are called fibers which consists of unit tangent vectors with the same origin x whose end-points describes circles. To express this fact E is said to be a circle bundle.

Our formulas (7)-(11) are valid for E, with special properties. Since e_3 is a normal vector, we have

$$(dx, e_3) = 0,$$

which gives, by the first equation of (8),

$$\omega_3 = 0.$$

By the first equation of (11), its exterior differentiation gives

$$\omega_1 \wedge \omega_{13} + \omega_2 \wedge \omega_{23} = 0.$$

Since M is immersed, ω_1 and ω_2 are linearly independent. By Cartan's lemma, we have

$$\begin{aligned} \omega_{13} &= a\omega_1 + b\omega_2, \\ \omega_{23} &= b\omega_1 + c\omega_2. \end{aligned} \tag{14}$$

In terms of the $\omega's$, the first and second fundamental forms are

$$\begin{aligned} I &= (dx, dx) = \omega_1^2 + \omega_2^2, \\ II &= -(dx, de_3) = \omega_1\omega_{13} + \omega_2\omega_{23} \\ &= a\omega_1^2 + 2b\omega_1\omega_2 + c\omega_2^2. \end{aligned} \tag{15}$$

Formally these forms are in E, but their definitions show that they are invariant under a rotation (13) of the vectors e_1, e_2 and are the pull-backs to E of well-defined forms of M. This method of using the analytical mechanism in E to study the geometry of M is fundamental in the theory of moving frames.

We will call ω_1, ω_2 the coframe dual of the frame e_1, e_2. Under the rotation (13), let ω_1', ω_2' by the new coframe. The first equation of (8) gives

$$dx = \omega_1 e_1 + \omega_2 e_2 = \omega_1' e_1' + \omega_2' e_2', \tag{16}$$

so that the coframe is transformed as follows:

$$\begin{aligned} \omega_1' &= \cos\tau\omega_1 + \sin\tau\omega_2, \\ \omega_2' &= -\sin\tau\omega_1 + \cos\tau\omega_2. \end{aligned} \tag{17}$$

In particular,

$$dA = \omega_1 \wedge \omega_2 \neq 0 \tag{18}$$

is invariant under the change (17). It is the element of area of M.

The principal curvatures k_1, k_2 are the roots of the equation

$$\begin{vmatrix} a-k & b \\ b & c-k \end{vmatrix} = 0. \tag{19}$$

Hence we have the following formulas for the mean curvature and the Gaussian curvature:

$$\begin{aligned} H &= \frac{1}{2}(a+c), \\ K &= ac - b^2. \end{aligned} \tag{20}$$

9.3 The connection form

The circle bundle E is of dimension 3, for x varies on the surface M and the fiber $\pi^{-1}(x)$ is one-dimensional.

Lemma 9.3.1. *The forms ω_1, ω_2, ω_{12} are uniquely defined on E and are linearly independent.*

Proof. The points of E are the unit tangent vectors xe_1. Since e_2 is completely determined by e_1, the first half of the equation (16) shows that ω_1, ω_2 are uniquely determined. □

Now the first equation of (11) gives

$$d\omega_1 = \omega_{12} \wedge \omega_2, \qquad d\omega_2 = \omega_1 \wedge \omega_{12}. \tag{21}$$

These equations completely determine ω_{12}. In fact, suppose that

$$d\omega_1 = \omega_{12}' \wedge \omega_2, \qquad d\omega_2 = \omega_1 \wedge \omega_{12}'.$$

Subtracting the respective equations, we get

$$(\omega_{12}' - \omega_{12}) \wedge \omega_2 = 0, \qquad \omega_1 \wedge (\omega_{12}' - \omega_{12}) = 0.$$

It follows that the difference $\omega_{12}' - \omega_{12}$ is a multiple of ω_2, and of ω_1. Hence it must be zero.

The form ω_{12} can be calculated as follows: Take a particular decomposition of the first fundamental form I as a sum of squares:

$$I = (\omega_1^0)^2 + (\omega_2^0)^2,$$

where ω_1^0, ω_2^0 are linear differential forms in u, v, and

$$\omega_1^0 \wedge \omega_2^0 = dA \neq 0.$$

Any two-form in u, v is a multiple of dA, and we can write

$$d\omega_1^0 = A\omega_1^0 \wedge \omega_2^0, \qquad d\omega_2^0 = B\omega_1^0 \wedge \omega_2^0, \tag{22}$$

where A, B are functions of u, v. Replacing ω_1, ω_2, ω_1', ω_2' in (16) and (17) by ω_1^0, ω_2^0, ω_1, ω_2, respectively, we have

$$\begin{aligned} \omega_1 &= \cos\tau\omega_1^0 + \sin\tau\omega_2^0, \\ \omega_2 &= -\sin\tau\omega_1^0 + \cos\tau\omega_2^0, \end{aligned} \tag{23}$$

and we can regard u, v, τ as local coordinates in E. By exterior differentiation we get

$$d\omega_1 = d\tau \wedge \omega_2 + (\cos\tau A + \sin\tau B)\omega_1 \wedge \omega_2,$$
$$d\omega_2 = -d\tau \wedge \omega_1 + (-\sin\tau A + \cos\tau B)\omega_1 \wedge \omega_2.$$

Since ω_{12} is uniquely determined by (21), we find

$$\omega_{12} = d\tau + A\omega_1^0 + B\omega_2^0. \tag{24}$$

It follows that

$$\omega_1 \wedge \omega_2 \wedge d\tau = \omega_1^0 \wedge \omega_2^0 \wedge d\tau \neq 0,$$

which proves the second statement of the lemma.

The form ω_{12} in E is called the connection form. Its importance cannot be over-exaggerated. It is completely determined by the equations (21), and hence by the first fundamental form I. By orthogonal projection from e_3, we define the absolute differential

$$\begin{aligned} De_1 &= (de_1)^{\perp} = \omega_{12}e_2, \\ De_2 &= (de_2)^{\perp} = -\omega_{12}e_1. \end{aligned} \tag{25}$$

This is called the Levi-Civita connection. It has the following geometrical interpretation: Consider a curve C on M defined by

$$u = u(t), \qquad v = v(t). \tag{26}$$

Along C, let

$$T = t^1 e_1 + t^2 e_2 \tag{27}$$

be a family of tangent vectors. By (25) we define

$$DT = (dt^1 - t^2\omega_{12})e_1 + (dt^2 + t^1\omega_{12})e_2. \tag{28}$$

The vectors T are said to be parallel along C in the sense of Levi-Civita if

$$DT = 0. \tag{29}$$

Under Levi-Civita parallelism the length of a vector remains constant, as it the scalar product of two vectors.

Geometrically the parallelism can be realized as follows: Let M^1 be the developable enveloped by the tangent planes to M along C. Since M and M^1 have the same normals along C and since the parallelism depends only on the normals, it is the same whether C is considered to be on M or M^1. On the other hand, the parallelism depends only on the first fundamental form and M^1 is flat, so that it becomes the ordinary parallelism when M^1 is "rolled" into a plane.

The exterior derivative of the connection form is, by the second equation of (11),

$$d\omega_{12} = -\omega_{13} \wedge \omega_{23}.$$

By (14) and (20) this can be written

$$d\omega_{12} = -K\omega_1 \wedge \omega_2. \tag{30}$$

This is an important relation. The right-hand side, being the negative of the Gaussian curvature times the element of area, is a two-form in M. Thus the connection form ω_{12} in the E, while its exterior derivative is in M. Such a property is called transgression.

An immediate consequence is Gauss's "Theorema egregium": The Gaussian curvature K depends only on the first fundamental form.

The Gauss-Bonnet formula can also be easily deduced from (30), but we will not give it here.

The other two equations of the second set of equations in (11) give

$$d\omega_{13} = \omega_{12} \wedge \omega_{23}, \qquad d\omega_{23} = \omega_{13} \wedge \omega_{12}. \tag{31}$$

These are the Codazzi equations. They can be interpreted as follows: Consider the diagram

$$\begin{array}{ccc} E & \xrightarrow{\tilde{g}} & E_0 \\ \pi \downarrow & & \downarrow \pi_0 \\ M & \xrightarrow{g} & S_0^2, \end{array} \tag{32}$$

where the right-hand column is the unit tangent bundle of the unit sphere S_0^2, g is the Gauss map, sending a point $x \in M$ to the unit normal vector e_3 at x, and $\tilde{g}$ is defined by parallelism in the space. S_0^2 being the unit sphere, our discussion on general surfaces applies. By the equation

$$de_3 = \omega_{31}e_1 + \omega_{32}e_2,$$

we see that ω_{31} ω_{32} is the pull-back, by $\tilde{g}^*$, of the coframe in E_0. By comparing the equations (21) and (31), we see that ω_{12} is the pull-back of the connection form of E_0. Thus we have the following interpretation of the Codazzi equations: The connection form ω_{12} is the pull-back under the Gauss map of the connection form in the unit tangent bundle of the unit sphere.

Since exterior differentiation commutes with the pull-back, it follows that $d\omega_{12}(= -K\omega_1 \wedge \omega_2$ by (30)) is the pull-back of $-dA_0$, where dA_0 is the element of area of S_0^2, which has Gaussian curvature $= +1$. This gives the well-known theorem

$$K = dA_0/dA. \tag{33}$$

Hence the Codazzi equation can be interpreted as a refinement of the formula (33).

9.4 The complex structure

The first fundamental form can be written

$$I = ds^2 = \omega_1^2 + \omega_2^2 = \varphi\overline{\varphi}, \tag{34}$$

where

$$\varphi = \omega_1 + i\omega_2 \tag{35}$$

and the bar denotes complex conjugation. Suppose the data be analytic. Under this assumption we can extend the variables to be complex numbers. By a well-known theorem on ordinary differential equations we can write

$$\varphi = \lambda d\zeta, \qquad \lambda \neq 0. \tag{36}$$

Let

$$\zeta = \xi + i\eta. \tag{37}$$

It follows that

$$ds^2 = |\lambda|^2(d\xi^2 + d\eta^2). \tag{38}$$

This means that our metric is conformal to the flat metric in the $\xi\eta$-plane, which is remarkable fact. By a theorem of Korn-Lichtenstein, this is also true under weaker smoothness condition. The proof is difficult, c.f. S. Chern, Proc AMS, 6 (1995), 771–782.

A complex-valued linear differential from φ, defined up to a factor, is said to defined a complex structure on the surface if

$$\varphi \wedge \overline{\varphi} \neq 0. \tag{39}$$

This means that at every point there is a coordinate ζ, by (36). If there is another coordinate ζ', we have

$$\varphi = \lambda d\zeta = \lambda' d\zeta',$$

which gives

$$d\zeta' = \frac{\lambda}{\lambda'} d\zeta,$$

i.e., ζ' is a holomorphic function of ζ. We call M a one-dimensional complex manifold or a Riemann surface. It has local coordinates which are complex numbers defined up to a holomorphic function. An oriented surface has an induced complex structure, given by φ.

The orientation is needed, because reversing it we have the complex structure defined by

$$\psi = \omega_2 + i\omega_1 = i(\omega_1 - i\omega_2) = i\overline{\varphi}.$$

We call it the conjugate complex structure, the local coordinate being $\overline{\zeta}$.

A complex-valued linear differential form can be written

$$\alpha = a\varphi + b\overline{\varphi}, \tag{40}$$

and this decomposition into the two summands is well-defined. We will write

$$\prod_{1,0} \alpha = a\varphi, \qquad \prod_{0,1} \alpha = b\overline{\varphi}, \tag{41}$$

and say that they are of bidegrees $(1,0)$ and $(0,1)$, respectively. In particular, if f is a complex-valued function, we define

$$\partial f = \prod_{1,0} df, \qquad \overline{\partial} f = \prod_{0,1} df. \tag{42}$$

In terms of the local coordinate z we have

$$\begin{aligned} df &= f_\xi d\xi + f_\eta d\eta \\ &= f_\zeta d\zeta + f_{\overline{\zeta}} d\overline{f}, \end{aligned} \tag{43}$$

so that

$$f_\zeta = \frac{1}{2}(f_\xi - i f_\eta), \qquad f_{\overline{\zeta}} = \frac{1}{2}(f_\zeta + i f_\eta). \tag{44}$$

Thus

$$\partial f = f_\zeta d\zeta, \qquad \overline{\partial} f = f_{\overline{\zeta}} d\overline{\zeta} \tag{45}$$

and ∂ (resp. $\overline{\partial}$) is exterior differentiation with respect to the variable ζ (resp. $\overline{\zeta}$).

More generally, if α is a complex-valued differential form, we dencte by $\partial\alpha/\partial\zeta$ (resp. $\partial\alpha/\partial\overline{\zeta}$) the operator $\partial/\partial\zeta$ (resp. $\partial/\partial\overline{\zeta}$) applied to its coefficients. Define

$$\partial\alpha = d\zeta \wedge \frac{\partial\alpha}{\partial\zeta}, \qquad \overline{\partial}\alpha = d\overline{\zeta} \wedge \frac{\partial\alpha}{\partial\zeta}. \tag{46}$$

The increase the bidegrees by $(1,0)$ and $(0,1)$, respectively, and we have

$$d = \partial + \overline{\partial}. \tag{47}$$

Since $d^2 = 0$, we have, by considering the bidegrees,

$$\partial^2 = \overline{\partial}^2 = 0, \qquad \partial\overline{\partial} + \overline{\partial}\partial = 0. \tag{48}$$

Differentiating (45), we find

$$d\overline{\partial} f = \partial\overline{\partial} f = \frac{1}{4}(f_{\xi\xi} + f_{\eta\eta}) d\zeta \wedge d\overline{\zeta}. \tag{49}$$

The function f is called holomorphic (resp. anti-holomorphic) if

$$\overline{\partial} f = 0 \quad (\text{resp. } \partial f = 0). \tag{50}$$

Also a form $\alpha = f(z)dz$ of bidegree $(1,0)$ is called holomorphic or abelian if

$$\overline{\partial}\alpha = 0.$$

This means that the coefficient $f(z)$ is a holomorphic function. The determination of the number of linearly independent holomorphic on a Riemann surface is called the Riemann-Roch problem.

A function is called harmonic if

$$\partial\overline{\partial}f = 0.$$

By (49) this is a generalization of the notion of a harmonic function in the euclidean plane. A holomorphic function is harmonic.

Let N be another surface with a complex structure defined by the form ψ and the local coordinate w. A mapping

$$f : M \to N$$

is called holomorphic (resp. anti-holomorphic) if it is given by a holomorphic function $w(\zeta)$ (resp. anti-holomorphic function $w(\overline{\zeta})$) or, what is the same, if the pull-back of ψ under f^* is a multiple of φ (resp. $\overline{\varphi}$).

10

Minimal Surfaces

10.1 General theorems

We consider the diagram

$$\begin{array}{c} E \\ \pi \downarrow \\ M \xrightarrow{x} E^3, \end{array} \tag{1}$$

where the bottom line is an immersed oriented surface in the three-dimensional euclidean space and E is the circle bundle of all frames $xe_1e_2e_3$, with $x \in M$ and e_3 the unit normal vector at x: π maps the frame $xe_1e_2e_3$ to its origin x.

From Chapter 9 we have the fundamental equations

$$\begin{aligned} dx &= \omega_1 e_1 + \omega_2 e_2, \\ de_1 &= \qquad\quad \omega_{12} e_2 + \omega_{13} e_3, \\ de_2 &= -\omega_{12} e_1 \qquad + \omega_{23} e_3, \\ de_3 &= -\omega_{13} e_1 \qquad - \omega_{23} e_3, \end{aligned} \tag{2}$$

$$\omega_{13} = a\omega_1 + b\omega_2, \quad \omega_{23} = b\omega_1 + c\omega_2, \tag{3}$$

$$d\omega_1 = \omega_{12} \wedge \omega_2, \quad d\omega_2 = \omega_1 \wedge \omega_{12}, \tag{4}$$

$$d\omega_{12} = -K\omega_{13} \wedge \omega_2, \tag{5}$$

$$d\omega_{13} = \omega_{12} \wedge \omega_{23}, \quad d\omega_{23} = \omega_{13} \wedge \omega_{12}. \tag{6}$$

In classical surface theory, the equations for de_1, de_2 in (2) are called the Gauss equations, the equation for de_3, the Weingarten equation, equation (5) also the Gauss equation, and the equation (6) the Codazzi equations.

In this chapter we shall study minimal surface characterized by the vanishing of the mean curvature:

$$H = \frac{1}{2}(a + c) = 0. \tag{7}$$

The surface M has a complex structure defined by $\varphi = \omega_1 + i\omega_2$. by the last equation of (2) the unit sphere S_0^2 has a complex structure $\omega_{13} + i\omega_{23}$. Under the Gauss map we have

$$\omega_{13} + i\omega_{23} = (a+ib)\overline{\varphi} + 2Hi\omega_2. \tag{8}$$

Hence we have the theorems:

Theorem 10.1.1. *The surface M is minimal surface if and only if the Gauss map is anti-holomorphic.*

We can write the first equation of (2) as

$$dx = \varphi \otimes \frac{1}{2}(e_1 - ie_2) + \overline{\varphi} \otimes \frac{1}{2}(e_1 + ie_2), \tag{9}$$

(generally we omit the tensor product sign $\otimes$) so that

$$2\partial x = \varphi \otimes (e_1 - ie_2). \tag{10}$$

By taking the exterior derivative of this question, using (2) and (4), we get

$$i\partial\overline{\partial} x = H\omega_1 \wedge \omega_2 \otimes e_3. \tag{11}$$

This gives the theorem:

Theorem 10.1.2. *The surface is a minimal surface, if and only if the coordinate vector (i.e., each of the coordinate functions) is a harmonic function (relative to the complex structure on the surface).*

We can write

$$\partial x = (x_1(\zeta), x_2(\zeta), x_3(\zeta))d\zeta, \tag{12}$$

where $x_i(\zeta)$, $1 \le i \le 3$, are holomorphic functions of ζ. By (10), the vector (x_1, x_2, x_3) is proportional to $e_1 - ie_2$. Since we have

$$(e_1 - ie_2, e_1 - ie_2) = 0,$$

we get

$$x_1^2 + x_2^2 + x_3^2 = 0. \tag{13}$$

If $[x_1, x_2, x_3]$ are considered as the homogeneous coordinates of the complex projective plane $P_2(\mathbb{C})$, equation (13) defines a conic Q. Its parametric equations can be written

$$\begin{aligned} x_1 &= -\frac{i}{2}(1-g^2)F, \\ x_2 &= \frac{1}{2}(1+g^2)F, \\ x_3 &= igF, \end{aligned} \tag{14}$$

where g is a parameter on Q taking values in $\mathbb{C} \cup \infty$, the Riemann sphere.

Since

$$dx_i = \partial x_i + \overline{\partial} x_i = 2\mathrm{Re}\partial x_i,$$

where

$$\begin{aligned} x_1 &= \mathrm{Re} \int -i(1 - g(\zeta)^2)F(\zeta)d\zeta, \\ x_2 &= \mathrm{Re} \int (1 + g(\zeta)^2)F(\zeta)d\zeta, \\ x_3 &= \mathrm{Re} \int 2ig(\zeta)F(\zeta)d\zeta. \end{aligned} \tag{15}$$

These are the Weierstrass formulas for a minimal surface, giving the surface in terms of two holomorphic functions $g(\zeta)$, $F(\zeta)$.

The function g defines a map

$$\tilde{g} : M \to Q, \tag{16}$$

which is essentially the Gauss map $g : M \to S_0^2$ define in (I, 32). For the define a map

$$\sigma : S_0^2 \to Q \tag{17}$$

as follows: Let $e_3 \in S_0^2$. Choose e_1, e_2, such that $\{e_1, e_2, e_3\}$ is a frame. Then e_1, e_2 are defined up to a rotation given by (chapter, (13)), so that $e_1 - ie_2$ is defined up to a factor of absolute value 1, and the map $\sigma(e_3) = [e_1 - ie_2]$ is well-defined. Conversely, given a point $x = (x_1, x_2, x_3)$ on Q such that $x \neq 0$, write $x = x' + ix''$, where x', x'' are real. Equation (13) implies

$$(x', x'') = 0, \qquad (x', x') = (x'', x'') > 0.$$

Set

$$e_1 = x'/(x', x')^{\frac{1}{2}} \qquad e_2 = x''/(x', x')^{\frac{1}{2}}.$$

Then x is the image under σ of $e_3 = e_1 \times e_2$. It is clear that σ is a diffeomorphism and $\tilde{g} = \sigma \circ g$.

We note the following corollaries of Theorem 10.1.2:

Corollary 10.1.3. *A minimal surface is real analytic.*

For it can be given as a graph $x_3 = F(x_1, x_2)$ expressing one coordinate of the space as a function of the two others. Since F is harmonic, it is analytic.

Corollary 10.1.4. *There is no compact minimal surface without boundary.*

Remark 10.1.5. *The Weierstrass formula is based on the fact that the curve $\frac{\partial z}{d\zeta}$ is a holomorphic curve. This idea has a generalization to the notion of a ∂-transform, which give rise to a new minimal surface from a given one. C.f. S. Chern and J. Wolfson, Harmonic maps of the two-sphere into a complex Grassmann manifold II, Annals of Math 125 (1987), 301–355.*

10.2 Examples

By (chapter 9, (6)), the equation for a non-parametric minimal surface $z = z(x,y)$ is

$$(1+z_y^2)z_{xx} - 2z_x z_y z_{xy} + (1+z_x^2)z_{yy} = 0.$$

This was first given by Lagrange in 1761. It is not easy even to give a local solution. The first examples of minimal surface, the helicoid and the catenoid, were given by Meusnier in 1776.

It is easy to give examples through the Weierstrass formula (15). The simplest ones are those for which $g(\zeta) = \zeta$. Putting $F = \frac{1}{2}w'''$, equation (15) can be integrated to give

$$\begin{aligned} x_1 &= \mathrm{Re}\, i\left(w - \zeta w' - \frac{1}{2}(1-\zeta^2)w''\right), \\ x_2 &= \mathrm{Re}\left(w - \zeta w' + \frac{1}{2}(1+\zeta^2)w''\right), \\ x_3 &= \mathrm{Re}(-i(w' - \zeta w'')). \end{aligned} \tag{18}$$

Perhaps the most remarkable minimal surface is one with $F = \text{const}$. Setting $F = +3i$ and $\zeta = \xi + i\eta$, we get its equation

$$\begin{aligned} x_1 &= \mathrm{Re}(3\zeta - \zeta^3) = 3\xi(1+\eta^2) - \xi^3, \\ x_2 &= \mathrm{Re}(i(3\zeta + \zeta^3)) = -3\eta(1+\xi^2) + \eta^3, \\ x_3 &= \mathrm{Re}(-3\zeta^2) = -3\xi^2 + 3\eta^2. \end{aligned} \tag{19}$$

This is called Enneper's surface. The following properties can be verified:

$$\begin{aligned} ds^2 &= 9(1+\xi^2+\eta^2)^2(d\xi^2 + d\eta^2), \\ K &= -\frac{4}{9}(1+\xi^2+\eta^2)^4. \end{aligned} \tag{20}$$

The tangent plane has the equation

$$2\xi x_1 + 2\eta x_2 + (1-\xi^2-\eta^2)x_3 = 3(\xi^2-\eta^2) + \xi^4 - \eta^4.$$

This gives Darboux's theorem: The Enneper surface is the envelope of the planes of symmetry of the points on the parabolas

$$\begin{aligned} p_1 &= 4\xi, \quad p_2 = 0, \quad p_3 = -2\xi^2 + 1, \\ q_1 &= 0, \quad q_2 = -4\eta, \quad q_3 = 2\eta^2 - 1. \end{aligned} \tag{21}$$

These parabolas lie in perpendicular planes and each vertex is the focus of the other.

Another example consists of the helicoidal minimal surfaces, obtained by setting

$$g(\zeta) = \zeta, \quad F(\zeta) = -\frac{ae^{i\alpha}}{\zeta^2}, \quad a > 0, \alpha \text{ real}. \tag{22}$$

Integrating (15), we get

$$\begin{aligned} \frac{x_1}{a} &= \operatorname{Re}\frac{e^{i\alpha}}{i}\left(\zeta + \frac{1}{\zeta}\right), \\ \frac{x_2}{a} &= -\operatorname{Re}^{i\alpha}\left(\zeta - \frac{1}{\zeta}\right), \\ \frac{x_3}{a} &= \operatorname{Re}\frac{2e^{i\alpha}}{i}(\log\zeta - 1). \end{aligned} \tag{23}$$

Putting $\log\zeta = u + iv$, gives

$$\begin{aligned} \frac{x_1}{2a} &= \sin\alpha \operatorname{ch} u \cos v + \cos\alpha \operatorname{sh} u \sin v, \\ \frac{x_2}{2a} &= \sin\alpha \operatorname{ch} u \sin v - \cos\alpha \operatorname{sh} u \cos v, \\ \frac{x_3}{2a} &= \sin\alpha u + \cos\alpha v. \end{aligned} \tag{24}$$

Two minimal surface are called adjoint if the function $F(\zeta)$ differ by a factor i. In (24) we ge a helicoid for $\alpha = 0$ and a catenoid for $\alpha = \frac{\pi}{2}$. They are adjoint minimal surfaces.

10.3 Bernstein -Osserman theorem

Osserman's Theorem. *Let $M \to E^3$ be a complete minimal surface which is not a plane, and let $g : M \to S_0^2$ be the Gauss map. Then the image $g(M)$ is dense in S_0^2, i.e., its closure $\overline{g(M)} = S_0^2$.*

Corollary 10.3.1 (Bernstein's Theorem). *Let*

$$z = z(x,y), \qquad \forall(x,y)$$

be a minimal surface in E^3 which has a one-one projection on the whole (x,y)-plane. Then $z(x,y)$ is a linear function.

For the surface is clearly complete and its Gauss map omits a half-sphere.

The proof of the theorem depends on a fundamental formula between the metric and the Gaussian curvature. We have, by (4),

$$d\varphi = i\varphi \wedge \omega_{12}, \tag{25}$$

where ω_{12}, the connection form, is uniquely defined by this equation and the property that it is a real form. (*Short proof*: If $d\varphi = i\varphi \wedge \omega'_{12}$, ω'_{12} real, we have

$\varphi \wedge (\omega_{12}' - \omega_{12}) = 0$, which implies $\omega_{12}' - \omega_{12} \equiv 0$, mod φ, and $\omega_{12}' - \omega_{12} = 0$, since it is real.)

We write (chapter 9, 36) as

$$\varphi = |\lambda| e^{i\tau} d\zeta, \tag{26}$$

so that ζ and τ are local coordinates in E, with τ real and $|\lambda|$ a function of ζ. Then

$$\begin{aligned} d\varphi &= d(\log(|\lambda| e^{i\tau})) \wedge \varphi \\ &= i\varphi \wedge (-d\tau + i\overline{\partial} \log |\lambda|) \\ &= i\varphi \wedge (-d\tau + i\overline{\partial} \log |\lambda| - i\partial \log |\lambda|). \end{aligned}$$

Hence

$$\omega_{12} = -d\tau + i(\overline{\partial} - \partial) \log |\lambda|. \tag{27}$$

Its exterior derivative is by (5),

$$2i\partial\overline{\partial} \log |\lambda| = -K\omega_1 \wedge \omega_2 \tag{28}$$

This fundamental formula expresses the "second derivative" of the metric in terms of curvature. It reminds us of Newton's second law of motion:

$$m\ddot{x} = F.$$

Force is curvature!

We rewrite the equation (14) of the conic Q in terms of homogeneous parameters $g : h$, thus

$$\begin{aligned} x_1 &= -\frac{i}{2}(h^2 - g^2)F, \\ x_2 &= \frac{1}{2}(h^2 + g^2)F, \\ x_3 &= ighF. \end{aligned} \tag{29}$$

The metric of M is

$$ds^2 = 2(\partial x, \overline{\partial} x), \tag{30}$$

which is, by (12) and (29), equal to

$$ds^2 = |F|^2 (g\overline{g} + h\overline{h})^2 d\zeta d\overline{\zeta}. \tag{31}$$

Lemma 10.3.2. *Let M be a complete minimal surface, which is not a plane. Suppose $u : M \to R$ is a function such that:*

1) $u \geq C)$ $(= Const.) > 0$;
2) $2i\partial\overline{\partial} \log u = K\omega_1 \wedge \omega_2$.

Then M is parabolic, i.e., its universal covering surface is conformally equivalent to the plane.

Proof. We suppose M to be simply connected, otherwise we take the universal covering surface of M. By condition 2) and (28) the function $G = \log(u\,\lambda|)$ is harmonic. Let $\tilde{G}$ be its conjugate harmonic function. Let

$$w = \int_0^\zeta \exp(G + i\tilde{G})d\zeta.$$

Then w is holomorphic and

$$|w'| = \exp G = u|\lambda| \neq 0,$$
$$|dw| = uds.$$

□

Let $w_0 = w(0)$. In a neighborhood of w_0 consider the inverse function $\zeta = \zeta^{-1}(w)$ and extend it analytically. Let Δ be a circular disc beyond which the extension is not possible. There must be a singularity w_1 on its boundary. It cannot be an algebraic branch point, because $w'(\zeta) \neq 0$. Therefore the line segment from w_0 to w_1 must correspond to an arc going to the boundary of M. Its length is

$$\int ds = \int \frac{|dw|}{u} \leq \frac{1}{C}|w_1 - w_0|.$$

This contradicts the completeness of M. Hence $\zeta^{-1}(w)$ established a biholomorphism between the w-plane and the universal covering surface of M.

Proof of Osserman's theorem. Let $\tilde{M}$ be the universal covering surface of M. It suffices to prove the theorem for $\tilde{M}$. Suppose $g(\tilde{M})$ omits a neighborhood of the point $g = 0$, $h = 1$ on the conic (29). Take

$$u = \frac{|g(\zeta)|^2}{|g(\zeta)|^2 + |h(\zeta)|^2}$$

By (28) and (31), this function satisfies the conditions of the lemma. It follows that $\tilde{M}$ is conformally the whole complex plane. But for $g \neq$ constant, this is not possible as otherwise there would be a non-constant entire function on the w-plane not taking values of a neighborhood. Hence the theorem is proved.

F. Xavier proved that under the assumption of Osserman's Theorem $S_0^2 - g(M)$ is a finite set of q points, with $q \leq 6$. Recently, Fugimoto proved the same statement with $q \leq 4$. In other words, if the Gauss map of a complete minimal surface omits five points or more on the unit sphere, it is a plane. By an example of Osserman, the exists a complete mininal surface whose Gauss map omits four-points so that Fugimoto's result is the best possible.

10.4 Inequality on Gaussian curvature

The considerations of the last section can be refined to give an estimate of the Gaussian curvature at the center of a minimal disk. The latter is

$$K = -a^2 - b^2 \leq 0.$$

The main tool is the Schwarz-Pick lemma: Let $w = w(\zeta)$ be a holomorphic function mapping the unit disc $|\zeta| < 1$ into the unit disc $|w| < 1$, such that $w(0) = 0$. Then

$$\left|\frac{dw}{d\zeta}\right|_{\zeta=0} \leq 1.$$

Pick gave a geometrical interpretation of this inequality: The disc $|\zeta| < 1$ has the hyperbolic metric

$$ds = \frac{|d\zeta|}{1-|\zeta|^2}$$

of curvature $= -4$. Thus the Schwarz-Pick lemma says that a holomorphic mapping shortens the hyperbolic distance. For the disc $|dw| < R$, the hyperbolic metric is

$$d\sigma = \frac{R|dw|}{R^2 - |w|^2}.$$

If the holomorphic mapping $w = f(\zeta)$ satisfies

$$|\zeta| < \rho, \quad |w| < R, \quad w_0 = f(0),$$

we will make use of the following infinitesimal form the Schwarz-Pick lemma:

$$\left|\frac{dw}{d\zeta}\right|_{\zeta=0} \leq \frac{1}{\rho R}(R^2 - |w_0|^2). \tag{32}$$

The map σ defined in (17) is essentially stereographic projection. In fact, let $e_3 = (v_1, v_2, v_3) \in S_0^2$, not the north pole N, so that $v_1^2 + v_2^2 + v_3^2 = 1$, $v_3 \neq 1$. It is easily verified that its image under σ is the point (14) with

$$g = -\frac{v_1 + iv_2}{1 - v_3} \tag{33}$$

(It suffices to check with this choice of g the vector (14) is orthogonal to e_3). We notice

$$1 + |g|^2 = \frac{2}{1 - v_3}, \tag{34}$$

which will be used later on.

Differentiating (33), we get

$$\begin{aligned} (1-v_3)^2 dg\overline{dg} &= dv_1^2 + dv_2^2 + dv_3^2 = \omega_{13}^2 + \omega_{23}^2 \\ &= (\omega_{13} + i\omega_{23})(\omega_{13} - i\omega_{23}), \end{aligned} \tag{35}$$

which, by (8) and (chapter 9, (36)), is equal to

$$(a^2+b^2)\varphi\overline{\varphi} = -K\varphi\overline{\varphi} = -K|\lambda|^2 d\zeta d\overline{\zeta}.$$

Since g is a holomorphic function of ζ, we have

$$\left|\frac{dg}{d\zeta}\right| = \sqrt{-K}\frac{|\lambda|}{1-v_3}. \tag{36}$$

Consider a disc on S_0^2 bounded by a latitude circle of spherical radius α about the north pole N. A point on the latitude has the coordinates $(\cos\theta\sin\alpha, \sin\theta\sin\alpha, \cos\alpha)$. By (33), $g = -\cot\frac{1}{2}\alpha e^{i\theta}$. Hence the complement of the disc is mapped by stereographic projection into a disc of radius, $\cot\frac{1}{2}\alpha$, in the complex g-plane.

Let Δ be a geodesic disc of radius s about a point ζ_0 on the minimal surface M. To apply the Schwarz-Pick lemma, we wish to estimate the size of a complex analytic disc contained in Δ.

The metric on S_0^2 is given by (35) and has Gaussian curvature $= 1$. Applying the formula (28) to this case, we get

$$2i\partial\overline{\partial}(1-v_3) = -\omega_{13}\wedge\omega_{23}.$$

Since the Gauss map is anti-holomorphic, this gives, by (34),

$$2i\partial\overline{\partial}\log(1+|g|^2) = -K\omega_1\wedge\omega_2. \tag{37}$$

Combining with (28), we get

$$i\partial\overline{\partial}\log(|\lambda|(1+|g|^2)^{-1}) = 0, \tag{38}$$

which means that

$$\log(|\lambda|(1+|g|^2)^{-1}) = G(\zeta) \tag{39}$$

is a harmonic function. Let $\tilde{G}(\zeta)$ be its conjugate harmonic function and let

$$\exp(G+i\tilde{G}) = f$$

Then $f : \Delta \to f$-plane is an unbranched covering. For the segment from $f(\zeta_0)$ to $f(\zeta_1)$, the length of its original curve is

$$\int_{\zeta_0}^{\zeta_1} |\lambda d\zeta| \leq (1+R^2)|f(\zeta_0) - f(\zeta_1)|.$$

Hence Δ contains a disc $|f(\zeta_0) - f| < s/(1+R^2) = \rho$, say.

To the function $g(\zeta(f))$ we now apply the Schwarz-Pick lemma (32). We can assume $\zeta_0 = 0$. We have, by (36),

$$\left|\frac{dg}{df}\right| = \left|\frac{dg}{d\zeta}\right| \bigg/ \left|\frac{df}{d\zeta}\right| = \frac{1}{2}\sqrt{-K}$$

By (32) this gives

$$\sqrt{-K_0} \leq \frac{2R}{\rho} = \frac{2}{s}\cot\frac{1}{2}\alpha\csc^2\frac{1}{2}\alpha, \tag{40}$$

where K_0 is the Gaussian curvature at the center of Δ. In other words, when the image under the Gauss map of a minimal disc is at a spherical distance $> \alpha$ from a point on the unit sphere, the Gaussian curvature at the center of the disc satisfies (40).

Inequality (40) gives an upper bound on the absolute value of the Gaussian curvature in terms of the size of a minimal disc. For example, if $\alpha = \frac{\pi}{2}$, where the Gauss map omits a hemisphere, (40) gives

$$\sqrt{-K_0} \leq \frac{4}{s}.$$

Letting $s \to \infty$, we have $K_0 = 0$. On a complete minimal surface any point can be taken as the center of an infinite disc. Hence the Gaussian curvature vanishes identically, and we have Bernstein's theorem.

An inequality of the type (40) was first given by E. Heinz in 1952. Inequality (40) is due to Osserman.

11

Pseudospherical Surface

11.1 General theorems

A pseudospherical surface is an immersed surface whose Gaussian curvature is a negative constant. Generally we suppose $K = -1$.

It has the following immediate properties:

1) There are no umbilics. For an umbilic is a point where the principal curvatures k_1, k_2 are equal, while $K = k_1 k_2$.

2) The asymptotic curves are real distinct.

If the surface is a graph $z = z(x,y)$, the condition $K = -1$ can be written

$$z_{xx}z_{yy} - z_{xy}^2 + (1 + z_x^2 + z_y^2)^2 = 0. \tag{1}$$

Hence we are dealing with the geometry of a Monge-Ampère equation.

The first and second fundamental forms are

$$\begin{aligned} I &= \omega_1^2 + \omega_2^2, \\ II &= k_1\omega_1^2 + k_2\omega_2^2, \quad k_1k_2 = -1, \end{aligned} \tag{2}$$

where e_1, e_2 are in principal directions. If we take the lines of curvature to be the u-curves and v-curves, respectively, we can write

$$\begin{aligned} \omega_1 &= Adu, \quad \omega_2 = Cdv, \\ \omega_{13} &= adu, \quad \omega_{23} = cdv. \end{aligned}$$

Differentiating the first equation, we get

$$d\omega_1 = A_v dv \wedge du, \quad d\omega_2 = C_u du \wedge dv.$$

Using (chapter 9, 21), we get by inspection

$$\omega_{12} = -\frac{A_v}{C}du + \frac{C_u}{A}dv.$$

Similarly, by the Codazzi equations (chapter 9, 31), we have

$$\omega_{12} = -\frac{a_v}{c}du + \frac{c_u}{a}dv.$$

Hence

$$\frac{A_v}{C} = \frac{a_v}{c}, \quad \frac{C_u}{A} = \frac{c_u}{a}.$$

Let

$$k_1 = \cot\psi, \quad k_2 = -\tan\psi, \tag{3}$$

so that 2ψ is the angle between the asymptotic directions. Then

$$k_1 = \frac{a}{A} = \cot\psi, \quad k_2 = \frac{c}{C} = -\tan\psi.$$

Differentiating the first equation with respective to v, we get

$$\psi_v = \tan\psi(\log A)_v$$

or

$$A = U(u)\sin\psi.$$

Similarly, we have

$$C = V(c)\cos\psi.$$

Hence we can choose the parameters u, v, such that

$$A = \sin\psi, \quad C = \cos\psi,$$
$$a = \cos\psi, \quad c = -\sin\psi.$$

In terms of these u, v, we have

$$I = \sin^2\psi du^2 + \cos^2\psi dv^2,$$
$$II = \sin\psi\cos\psi(du^2 - dv^2), \tag{4}$$

$$\omega_{12} = -\psi_v du - \psi_u dv. \tag{5}$$

By the Gauss equation (chapter 9, 30), we get

$$\psi_{uu} - \psi_{vv} = -\sin\psi\cos\psi. \tag{6}$$

We may also introduce the coordinates p, q by

$$2p = u + v, \qquad 2q = -u + v. \tag{7}$$

Then

$$
\begin{aligned}
I &= dp^2 + 2\cos 2\psi dpdq + dq^2, \\
II &= 2\sin 2\psi dpdq.
\end{aligned} \tag{8}
$$

The second equation shows that the (p,q)-curves form the asymptotic net. The first equation shows that it is a fishing net and is called a Tchebychef net.

Equation (6) becomes

$$(2\psi)_{pq} = \sin 2\psi. \tag{9}$$

Either (6) or (9) is called the sine-Gordon equation. Thus a pseudospherical surface gives rise to a solution of the sine-Gordon equation. Conversely, and such solution $\psi(u,v)$ determines the forms ω_1, ω_2, ω_{13}, ω_{23}, ω_{12} satisfying all the Maurer-Cartan equations. They define a pseudospherical surface up to motion.

The most famous examples of a p. s. s. is Beltrami's pseudosphere, which we will describe below.

11.2 Bäcklund's theorem

Let the surfaces M and M^* be diffeomorphic such that the line joining corresponding points $x \in M$ and $x^* \in M^*$ is tangent to both surfaces. Let ν (resp. ν^*) be the unit normal vector to M (resp. M^*) at x (resp. x^*). Let

$$\overline{xx^*} = r, \quad (\nu, \nu^*) = \cos\theta, \tag{10}$$

so that r is the distance between x, xx^* and θ is the angle between the normals at x, x^*.

Bäcklund's Theorem 2.1. If $r = \text{const}$, $\theta = \text{const}$, $(\neq 0, \pi)$, then the Gaussian curvatures of M and M^* are

$$K = K^* = -\frac{\sin^2\theta}{r^2} \tag{11}$$

We choose the frames $xe_1e_2e_3$ and $x^*e_1^*e_2^*e_3^*$ so that $e_1^* = e_1$ is the unit vector in the direction $\overrightarrow{xx}^*$, and $e_3 = \nu$, $e_3^* = \nu^*$. Let

$$
\begin{aligned}
dx &= \varphi_1 e_1 + \varphi_2 e_2, \\
de_1 &= \qquad\quad \varphi_{12} e_2 + \varphi_{13} e_3, \\
de_2 &= -\varphi_{12} e_1 \qquad\quad + \varphi_{23} e_3, \\
de_3 &= -\varphi_{13} e_1 - \varphi_{23} e_2,
\end{aligned} \tag{12}
$$

and similar equations with asterisks. Equations (10) can be written

$$
\begin{aligned}
x^* &= x + re_1, \\
e_2^* &= \cos\theta e_2 + \sin\theta e_3, \\
e_3^* &= -\sin\theta e_2 + \cos\theta e_3.
\end{aligned} \tag{13}
$$

To prove Bäcklund's theorem, suppose the hypotheses are satisfied. Differentiating the first equation of (13), we have

$$dx^* = \varphi_1^* e_1 + \varphi_2^* e_2^* = \varphi_1 e_1 + (\varphi_2 + r\varphi_{12}) e_2 + r\varphi_{13} e_3,$$

which gives, by (13),

$$\begin{aligned} \varphi_1^* &= \varphi_1, \\ \varphi_2^* \cos\theta &= \varphi_2 + r\varphi_{12}, \\ \varphi_2^* \sin\theta &= r\varphi_{13}. \end{aligned}$$

The last two equations can be written

$$\varphi_2 + r\varphi_{12} - r\cot\theta\,\varphi_{13} = 0, \tag{14}$$

and

$$\varphi_2^* = \frac{r}{\sin\theta}\varphi_{13}$$

or

$$\varphi_2^* = R\varphi_{13},$$

where

$$r = R\sin\theta.$$

On the other hand, differentiating the equation $e_1^* = e_1$, we get

$$\begin{aligned} \varphi_{12}^* &= \cos\varphi_{12} + \sin\theta\,\varphi_{13}, \\ \varphi_{13}^* &= -\sin\theta\,\varphi_{12} + \cos\theta\,\varphi_{13} = R^{-1}\varphi_2, \end{aligned}$$

by (14). Differentiating the second equation of (13), we have

$$\varphi_{23}^* = \varphi_{23}.$$

Combining the above equations, we have

$$\begin{aligned} \varphi_1^* \wedge \varphi_2^* &= R\varphi_1 \wedge \varphi_{13}, \\ R\varphi_{13}^* \wedge \varphi_{23}^* &= R^2 K^* \varphi_1^* \wedge \varphi_2^* = R\varphi_2 \wedge \varphi_{23}. \end{aligned}$$

The sum of the right-hand sides is zero. Hence the same is true of the left-hand sides and this gives

$$K^* = -R^{-2}.$$

By symmetry, we also have $K = -R^{-2}$. Thus Bäcklund's theorem is proved.

Given M, to find the second surface M^* so that the pair M, M^* satisfies the conditions of Bäcklund's theorem, it suffices to have a vector field xe_1 on M. For the lines through x, $x \in M$, and in the direction of e_1 form a line congruence (∞^2 lines) which consists of the common tangent lines of M and M^* and M^* is the other "focal surface" enveloped by these lines. The vector field xe_1 satisfies the condition (14), which can be written

$$(dx, v, e_1) + r(de_1, v, e_1) - r\cot\theta(de_1, v) = 0. \tag{14a}$$

With e_1 as the unknown unit tangent vector, this is a total differential equation. It satisfies the condition of Frobenius' theorem. In fact, the exterior derivative of the left-hand side of (14) is

$$\varphi_1 \wedge \varphi_{12} + r\varphi_{13} \wedge \varphi_{32} - r\cot\theta\,\varphi_{12} \wedge \varphi_{23},$$

which is seen to be zero modulo the equation (14). This means that with given r, θ satisfying $r^2K = -\sin^2\theta$, the unit tangent bundle $E \longrightarrow M$ has a foliation whose leaves correspond to the vector fields of M which have the property of giving rise to a Bäcklund pair of surfaces. We will call such a vector field a Bäcklund field. A Bäcklund field is completely determined by an initial value, i.e., the vector at a point x_0.

As stated at the beginning of this Chapter, we will suppose $K = -1$ for simplicity, so that $R = 1$, $r = \sin\theta$. We will first consider the case $\theta = \frac{\pi}{2}$. In this case $\overline{xx^*} = 1$, $v \perp v^*$. Denote by S_x^1 the unit circle centered at $x \in M$ and lying on the tangent plane T_x to M at x. Then M^* is an orthogonal trajectory of this family of circles. Stated in another way, we have the theorem of Ribaucour: Let M be a surface with $K = -1$. In the tangent plane T_x, $x \in M$, let S_x^1 be the unit circle with center at x. Then the circles S_x^1 have an orthogonal trajectory which is a surface with Gaussian curvature $= -1$.

Proposition 11.2.1. *Under the conditions of Bäcklund's theorem, the lines of curvature on M and M^* correspond, as do their asymptotic curves.*

The proposition follows from the facts that the lines of curvatures and the asymptotic curves are defined respectively by the equations (ordinary multiplication)

$$\begin{aligned} \varphi_1\varphi_{23} - \varphi_2\varphi_{13} &= 0, \\ \varphi_1\varphi_{13} + \varphi_2\varphi_{23} &= 0. \end{aligned} \tag{15}$$

The above formulas give the proposition through substitutions.

A family of lines depending on two parameters is called a line congruence. Generally it consists of the common tangent lines of two surfaces, the focal surfaces. The congruence is called a W-congruence if the asymptotic curves of the focal surfaces correspond. It is called pseudospherical if the condition of Bäcklund's theorem are fulfilled. Bäcklund's theorem says that the focal surfaces of a pseudospherical congruence are pseudospherical surfaces. Proposition 11.2.1 says that a pseudospherical congruence is a W-congruence.

We will write equation (14) or (14a) in terms of the parameters u, v, or p, q introduced in §1. Let a_1, a_2 be the unit vectors tangent to be the u-curves and v-curves respectively. The vector e_1 is determined by the angle α it makes with a_1. Write

$$\begin{aligned} e_1 &= \cos\alpha a_1 + \sin\alpha a_2, \\ e_2 &= -\sin\alpha a_1 + \cos\alpha a_2. \end{aligned} \tag{16}$$

By §1 we have

$$\begin{aligned} &\omega_1 = \sin\psi du, \quad \omega_2 = \cos\psi dv, \\ &\omega_{13} = \cos\psi du, \quad \omega_{23} = -\sin\psi dv, \\ &\omega_{12} = -\psi_v du - \psi_u dv. \end{aligned} \tag{17}$$

By (16) we find

$$\begin{aligned} \varphi_2 &= -\sin\alpha\omega_1 + \cos\alpha\omega_2, \\ \varphi_{12} &= d\alpha + \omega_{12}, \\ \varphi_{13} &= \cos\alpha\omega_{13} + \sin\alpha\omega_{23}. \end{aligned} \tag{18}$$

Hence (14) becomes

$$\begin{aligned} d\alpha + \omega_{12} =& (\cot\theta\cos\alpha\cos\psi + R^{-1}\csc\theta\sin\alpha\sin\psi)du \\ &+ (-\cot\theta\sin\alpha\sin\psi - R^{-1}\csc\theta\cos\alpha\cos\psi)dv. \end{aligned} \tag{19}$$

In terms of the asymptotic parameters p, q it becomes

$$\begin{aligned} d\alpha =& \psi_p dp - \psi_q dq + (\cot\theta - R^{-1}\csc\theta)\cos(\alpha+\psi)dp \\ &- (\cot\theta + R^{-1}\csc\theta)\cos(\alpha-\psi)dq. \end{aligned} \tag{20}$$

The last equation can be written

$$\begin{aligned} (\alpha-\psi)_p &= (\cot\theta - R^{-1}\csc\theta)\cos(\alpha+\psi), \\ (\alpha+\psi)_p &= -(\cot\theta + R^{-1}\csc\theta)\cos(\alpha-\psi). \end{aligned} \tag{21}$$

These are the formulas for the Bäcklund transformation usually found in the literature.

As an application we wish to derive Dini's pseudospherical surface. We suppose $R = 1$, or $K = -1$. The transformation formulas are

$$\begin{aligned} &\varphi_1^* = \varphi_1, \quad \varphi_2^* = \varphi_{13}, \\ &\varphi_{13}^* = \varphi_2, \quad \varphi_{23}^* = \varphi_{23}, \\ &\varphi_{12}^* = \cos\theta\varphi_{12} + \sin\theta\varphi_{13}, \end{aligned} \tag{22}$$

with

$$\varphi_2 + \sin\theta\varphi_{12} - \cos\theta\varphi_{13} = 0. \tag{23}$$

Instead of an immersed surface M we take the elements of contact whose points lie on line L and whose planes pass through L. (An element of contact is a point and an oriented plan through it.) To describe this family of elements of contact we take a fixed frame $0e_1^0e_2^0e_3^0$ and suppose L be the line $0e_2^0$. Let the point be $x = ue_2^0$ and the plane be spanned by $e_1'e_2^0$, where

$$e_1' = \cos ve_1^0 + \sin ve_3^0,$$
$$e_3' = -\sin ve_1^0 + \cos ve_3^0,$$

so that $0e_1'e_2^0e_3'$ is a frame and u, v are coordinates in the family. A unit vector in the plane $xe_1'e_2^0$ can be supposed to be

$$e_1 = \cos we_1' + \sin we_2^0,$$

and we set

$$e_2 = -\sin we_1' + \cos we_2^0,$$

so that $xe_1e_2e_3$, $e_3 = e_1 \times e_2 = e_2'$, is a frame. Of this family of frames, we calculate the $\varphi's$ in (12), obtaining

$$\varphi_1 = \sin w du, \quad \varphi_2 = \cos w du$$
$$\varphi_{12} = dw, \quad \varphi_{13} = \cos w dv, \quad \varphi_{23} = -\sin w dv.$$

Hence equation (23) becomes

$$\csc\theta du - \cot\theta dv + \frac{dw}{\cos w} = 0.$$

This equation determines w as a function of u, v. To integrate it we set

$$\tau = u\csc\theta - v\cot\theta.$$

Then the equation becomes

$$d\tau + \frac{dw}{\cos w} = 0.$$

This equation can be integrated, giving, by a suitable choice of the constant of integration,

$$\cos w = -\text{sech } \tau, \quad \sin w = -\tanh \tau.$$

In terms of the fixed frame $e_1^0e_2^0e_3^0$ we have

$$e_1 = -\text{sech } \tau \cos ve_1^0 - \tanh \tau e_2^0 - \text{sech } \tau \sin ve_3^0$$

and

$$\begin{aligned} x^* &= x + \sin\theta e_1 \\ &= -\sin\theta \text{sech } \tau \cos ve_1^0 + (u - \sin\theta \tanh \tau)e_2^0 - \sin\theta \text{sech } \tau \sin ve_3^0. \end{aligned}$$

This defines Dini's p.s.s. When $\theta = \frac{\pi}{2}$, it becomes

$$x^* = -\text{sech}\, u \cos v e_1^0 + (u - \tanh u) e_2^0) - \text{sech}\, u \sin v e_3^0,$$

which is the famous Beltrami pseudosphere.

This example has an important implication; it leads to the notion of a generalized surface. The elements of contact of E^3 from a five-dimensional space Y. It can be identified with the space whose elements consist of a point x and a unit vector e_3, the plane being then the plane orthogonal to e_3. In Y there is a linear differential form $(dx, e_3) = \varphi_3$, which has the property

$$\begin{aligned}\varphi_3 \wedge (d\varphi_3)^2 &= \varphi_3 \wedge (\varphi_1 \wedge \varphi_{13} + \varphi_2 \wedge \varphi_{23})^2 \\ &= -2\varphi_1 \wedge \varphi_2 \wedge \varphi_3 \wedge \varphi_{13} \wedge \varphi_{23} \neq 0.\end{aligned}$$

Y is usually called the projectivized cotangent bundle over E^3.

A surface in Y annihilating φ_3 is called a Legendre surface. An example is the above surface M which consists of the elements of contact whose points lie on a line L and whose planes pass through L. Also the Dini surface, and hence the Beltrami pseudosphere, is a Legendre surface in Y without any singularity. Only its projection in E^3 gives rise to singularities.

Made in the USA
Monee, IL
01 March 2026

45213746R00142